The Art of Timing Closure

Khosrow Golshan

The Art of Timing Closure

Advanced ASIC Design Implementation

Khosrow Golshan
Laguna Beach, CA, USA

ISBN 978-3-030-49638-8   ISBN 978-3-030-49636-4   (eBook)
https://doi.org/10.1007/978-3-030-49636-4

This Springer imprint is published by the registered company Springer Nature Switzerland AG
The registered company address is: Gewerbestrasse 11, 6330 Cham, Switzerland

To those victims of the COVID-19 virus pandemic and the loved ones they left behind. Their tragedies make us all realize the importance of family.
In addition, I give sincere gratitude to the medical professionals and all those who work in essential businesses as they are the first responders who risk their lives to save others.

Foreword

ASIC design signoff and closure have become a very challenging process given the more stringent requirements for advanced technology nodes.

Pressure for quicker time-to-market in many industries such as Internet-of-Things (IoT), smart phones, artificial intelligence, speech and handwriting recognition, virtual reality, and 3D visualization is increasing. Furthermore, the sheer complexity of ASIC design requirements in area, power, and timing demands a thorough methodology for signoff and design closure.

Fortunately, we have a great book in hand that addresses the very issues encountered in the everyday design of complex ASIC systems. This book offers a detailed walkthrough of each step of the design process—from structure of the design all the way to design signoff and closure.

I really appreciate the detailed explanation of real problems and remedies throughout the implementation flow, specifically in the areas of timing closure and signoff methodologies.

This is a great hands-on, detailed, and technical reference book which walks through developing implementation methodology. I have rarely seen any book specifically addressing timing closure flow steps with ample explanation.

In addition, having the automation scripts is a great plus to the reader, as they enrich their tools with the concepts illustrated in the book.

Synopsys, Inc.   Shahin Golshan
Mountain View, CA, USA

Trademarks

Verilog, Encounter Design Integration, NanoRoute, Encounter System, Innovus, Conformal, and CPF are registered trademarks of Cadence Design Systems, Inc.

Liberty, Prime Time, Formality, UPF, and ICC are registered trademarks of Synopsys Systems, Inc.

SDF and SPEF are trademarks of Open Verilog International.

All other brand or product names mentioned in this document are trademarks or registered trademarks of their respective companies or organizations.

Disclaimer: The information contained in this manuscript is an original work of the author and intended for informational purposes only. The author disclaims any responsibility or liability associated with the content, the use, or any implementations created based upon such content.

Preface

The goal of this book is to provide the essential steps required for advanced design implementation of Application Specific Integrated Circuits (ASIC) regarding timing closures. It is the intention that the book presents a hands-on approach in solving the most challenging parts of design implementation.

Since 1987, a transistor gate length was about 3 μm, and today after almost three decades, we are dealing with transistors with gate length of 7 nm and below and increasingly complex ASIC design. Thus, the signoff process in regard to physical design and Static Timing Analysis (STA) and time-to-market has created a challenge. The following illustration shows the trend of lowering the size of transistors and increasing design complexity with modern Integrated Circuit (IC) process variations.

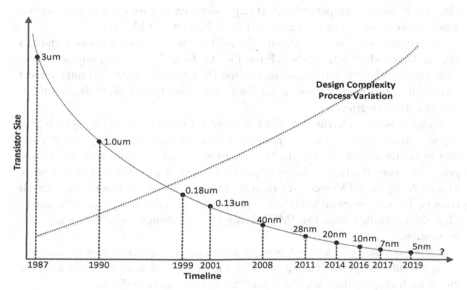

It is important to note that the transistor's operating voltage is proportional to its gate length (i.e., at 3 µm the transistor's operating voltage would be 3.0 V, 0.18 µm at 1.8 V and so on). At lower nodes, there are millions more transistors and the power consumption is a critical matrix (i.e., smaller transistors have more leakage and higher operating frequency has more dynamic power).

It should be noted that the 5 nm node was the first built using Extreme Ultraviolet (EUV) lithography, and it used 13.5 nm wavelength to produce extremely fine patterns on the silicon. Since there are not many numbers left between five and zero, what is next?

If there is one truism about Moore's law, is that transistors have been made smaller and smaller over past recent decades. Scientists and engineers have taken the challenge of creating a transistor that is one-atom thick. But for now, silicon remains the mainstream.

Regarding physical design and timing analysis, the initial timing analysis on larger process nodes, such as 180 and 130 nm, was concerned mostly with operation at worst-case and best-case conditions. The distance between adjacent routing tracks was such that coupling capacitances were marginalized by ground and pin capacitance. Because of that, the potential issues with crosstalk and noise effects were neglected, and the design margin for analyzing the crosstalk and noise would be increased.

Starting at 90 nm, and even more prominently at 65 nm, an increase in coupling capacitance due to narrower routing spaces and the thickened metal segment profiles resulted in crosstalk effects becoming a significant concern. Another area of concern was different transistor's voltages for speed (Low-VT) and leakage power reduction (High-VT). To mitigate these issues, two additional corners were added. One was to address temperature effect (e.g., worst-case at −40 °C) and the second was to analyze the design leakage power (e.g., best-case at 125 °C).

As process nodes became smaller (i.e., 40 nm and below), process variations became larger which affected both Front-End-Of-Line (FEOL), i.e., implant, diffusion, and poly layers, and Back-End-Of-line (BEOL), i.e., metal and interconnect layers. These process variations required additional timing analysis during advanced design implementation.

Today, however, increased product demand for more functionality with higher application frequencies at lower operating power has required a change in the complexity of the design. As designs become more complex, with many modes and process corners, the time required to perform a typical physical design and Static Timing Analysis (STA) increases in order to cover all design modes and process corners. Thus, to cover all today's design modes and corners efficiently, the use of Multi-Mode Multi-Corner (MMMC) during physical design and timing analysis is imperative.

The Art of Timing Closure—Advanced ASIC Design Implementation is formatted using a hands-on approach using MMMC during physical design and STA. At the end of each chapter, there are corresponding *Pearl* scripts as reference.

The scripts in this book are based on **Cadence® Encounter System™**. However, if the reader uses a different EDA tool, that tool's commands are similar to those shown in this book.

The topics covered are as follows:

- Data Structures
- Multi-Mode Multi-Corner Analysis
- Design Constraints
- Floorplan and Timing
- Placement and Timing
- Clock Tree Synthesis
- Final Route and Timing
- Design Signoff

Rather than going into lengthy technical depths, the emphasis has been placed on short, clear descriptions implemented by references to authoritative manuscripts. It is the goal of this book to capture the essence of physical design and timing analysis and to show the reader that physical design and timing analysis engineering should be viewed as a single area of expertise.

Laguna Beach, CA, USA   Khosrow Golshan
April 2020

Acknowledgments

Having an idea and turning it into a book is as hard as it sounds. The experience is both internally challenging and rewarding.

First and foremost, I want to thank Maury Golshan, my wife, for her editorial experience, dedication, and patience which helped make this happen. She spent a considerable amount of time and effort in proofreading and revising the book. Because of this, clarity and consistency of this manuscript was significantly improved. Without her dedication, it would be almost impossible to have completed this book.

In addition, I would like to express my gratitude to several individuals who contributed their time and effort towards this manuscript. Especially, Charles B. Glaser, Editorial Director at Springer, and his colleagues for their assistance and support in publishing this manuscript.

Khosrow Golshan

Contents

About the Author

Khosrow Golshan was Division Director at Conexant System Inc. and Technical Director at Synaptics Inc. while managing and directing worldwide ASIC design implementation and standard cell and I/O library development for various silicon process nodes. Prior to that he was Group Technical Staff at Texas Instrument's R&D and Process Development Laboratory responsible for processing silicon test-chip design and digital/mixed-signal ASIC development. He has over 20 years of experience in ASIC design implementation methodology, flow development, and digital ASIC libraries design.

He is the author of *Physical Design Essentials—An ASIC Design Implementation Perspective*. In addition, he has published many technical articles and has held several US patents.

The author has earned advanced degrees in the areas of Electrical Engineering (West Coast University, Los Angeles, CA. Engineering Dept.), Applied Mathematics (Southern Methodist University, Dallas, TX. Mathematics Dept.), and a Bachelor of Science in Electronic Engineering (DeVry University, Dallas, TX. Engineering Dept.). He is also an IEEE life member.

Chapter 1
Data Structures and Views

It is better to have 100 functions operate on one data structure than to have 10 functions operate on 10 data structures.
Alan Perlis

Data structures are a specialized format for organizing, processing, retrieving, and storing data. While there are several basic and advanced structure types, any data structure is designed to arrange data to suit a specific purpose so that it can be accessed and worked with in appropriate ways.

For advanced design implementation, data structures impact design quality and efficiency of the design flow. They are arranged as directories and sub-directories which contain various application-specific integrated circuit (ASIC) design data and views (functional and operational) in the form of binary and/or textual. These data are used during different stages of ASIC design implementation flow such as synthesis, place and route, timing analysis, etc.

Operational views are EDA tools-related data that are used during ASIC design and implementation. Most often they are in the form of binary databases. These tools are either provided by EDA providers or developed internally. The functional views are data files that are used during ASIC design and implementation and integration.

Both functional and operational views, either internal or external, must have an aggressive quality control (QC) flow. For that, one could use standard ASIC design flow to ensure their accuracy and quality.

It is important to note that quality control is used to verify that deliverables are of acceptable quality and that they are complete and correct. Examples of quality control activities include inspection, deliverable peer reviews, and the testing process. Quality control is about adherence to requirements. Failure to have a well-thought-out QC flow and standards could impact the end results for a manufactured ASIC product.

© Springer Nature Switzerland AG 2020
K. Golshan, *The Art of Timing Closure*,
https://doi.org/10.1007/978-3-030-49636-4_1

1.1   Data Structures

There are two types of data structures. One is common data structure (i.e., a directory or repository) which contains all views used by all ASIC projects. All its view files need to be an atomic collection in order to ensure data integrity. The other type of data structure is project data structure, which is a directory wherein the actual design is implemented and stored.

Both common and project data structures need to have data management. Data management is an administrative process that includes acquiring, validating, storing, protecting, and processing required data to ensure the accessibility, reliability, and timeliness of the data for its users. In general, common data structure is managed by the system administrator, and project data structure is managed by software known as version control system (VCS).

VCS manages multiple versions of computer files and programs. VCS provides two primary data management capabilities. The first allows users to lock files so they can only be edited by one person at a time. The second tracks changes to files.

For example, if there is only one user editing a document, there is no need to lock a file for editing. However, if a team of design implementation engineers are working on a project, it is important that no two people are editing the same file at the same time. If this happens, it is possible for one person to accidentally overwrite the changes made by someone else.

For this reason, VCS allows users to check out files for editing. When a file has been checked out from a shared file server, it cannot be edited by other users. Once the user finishes editing the file, the user saves the changes and checks in the file so that other users can then edit the file.

VCS also allows users to track changes to files by means of a tag. This type of version control is often used in design implementation development (the original intent of VCS was for software development) and is also known as source control or revision control.

Popular versions of control systems like subversion (SVN) and VCS allow ASIC and implementation designers to save incremental versions of their programs and source code, such as RTL files, during the design development process. This provides the capability to rollback to an earlier version of the program or source codes if necessary. For example, if bugs are found in a new version of RTL code, the design engineer can review the previous version when debugging the RTL code.

1.2   Common Data Structures

It is imperative to have a well-thought-out common data structure that it is used by multiple ASIC projects.

An example of a common data structure, as shown in Fig. 1.1, consists of five subdirectories: intellectual property (IP), libraries, tools, repository, and templates. The

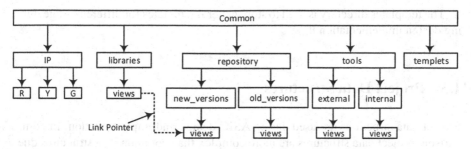

Fig. 1.1 Common data structure example

tools directory has two sub-directories, one for storing external software provided by EDA providers and another for internally developed software and programs.

The IP directory is used for company internal IP development and has three sub-directories. The red (R) is used for IP development; the yellow (Y) contains developed IP that has not gone through the full QC process and can be used as a risk factor; the green (G) contains the IP that has completed all QC processes, has been version tagged, and can be used by a given project.

The repository directory has two sub-directories—the new version and old version—and is used for storing functional views, either external or internal data such as standard cell, I/O (input and output) pads, and externally provided IP libraries.

Functional views that are stored in the repository directory are collections of the physical layouts, abstract views, timing models, simulation models, and transistor-level circuit description data. For example, the most common library views for an ASIC design are:

- lib_file: Timing files for Static Timing Analysis (STA)
- cdl_file: Circuit description language for physical verification
- spi_mod: Spice models for transistor-level circuit simulation
- tech_file: Technology files for various EDA tools
- lef_file: Library Exchange Format (LEF) for abstract files, standard cells, IOs and IPs for place and route tools

The library directory needs to have a link pointer to the latest version of the repository area and not the actual repository views. The reason for this is to allow users to create scripts and programs without changing them once a new version is available in the repository directory.

The tools directory is used for storing EDA external and internal software such as:

- Simulation (analog, mixed signal, and digital)
- Timing (STA)
- Memory compiler (static random access, read-only memory)
- Design integration (place and route, layout, power analysis)
- Verification (physical and formal analysis)
- Internal (productivity scripts and programs)

The templates directory is used to store general templates for different usage during design implementation flow.

1.3   Project Data Structures

Project data structures are used during ASIC design and implementation. In comparison, project data structures are more complex than common data structures due to their usage among different ASIC design and implementation engineers. It is important that all project data structures be the same for all design projects.

Although many companies that are involved in ASIC design projects have their own style of creating project spaces, having consistent functions and views among all projects is a process through which one can collect and organize data and maintain performance operation in the most effective way.

In addition, it is important to have the same naming convention for functions and views. One of the advantages of having a naming convention is that it allows users to find data and its location with ease. It is also useful to design implementation script developers when creating functional scripts covering all related projects.

In order to create a project data structure for a project, one basically needs to perform the following three steps:

- Take input
- Process it
- Provide output

The first step is to take user input. The input can be in any form (e.g., a simple text file containing the information about the project such as project name, the technology node, the location of design views, etc.).

The second step requires some sort of programming script – for example, *Perl*, *Tcl* (Tool Command Language), or *make* (*make* reads in rules from user-created *Makefile*).

The third step is to provide output (e.g., creating project directories, linking to central view libraries, etc.).

To make this process efficient, all three steps need to be optimized.

Figure 1.2 shows an example of project data structures for an ASIC design project called *Moonwalk*. The project has two sub-directories, design and implementation. Both sub-directories have link pointers to the common data structure.

Having data structure as such will allow users to create ASIC design and implementation-related scripts (such as synthesis), thus minimizing editorial errors later in the design and implementation flow. Design and implementation directories should be created by the system administrator once a new project space has been requested.

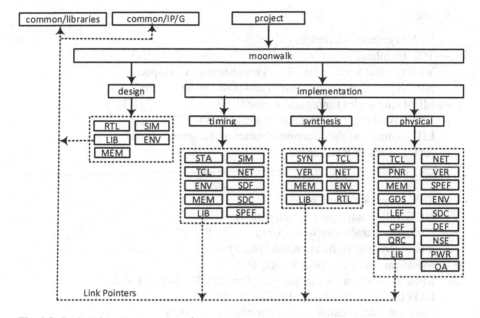

Fig. 1.2 Project data structure example

Design directory is used for ASIC product design development and has five sub-directories:

- RTL (RTL development workspace)
- SIM (RTL simulation workplace)
- LIB (pointers to the common directory and IP directories)
- ENV (design-related environmental files)
- MEM (memories timing and netlist files)

Implementation directory has three sub-directories, synthesis, physical, and timing. Each of its sub-directories is used in different stages of ASIC design implementation flow. The description of these three sub-directories' contents is as follows:

- Timing:

  - STA (STA workspace)
  - SIM (gate-level simulation workspace)
  - TCL (static timing and simulation analysis scripts)
  - NET (post-layout netlist)
  - ENV (gate-level simulation and STA environmental files)
  - SDF (Standard Delay Format for gate-level simulation)
  - MEM (memories timing and netlist files)
  - SDC (design constraints files)
  - SPEF (extraction files in Standard Parasitic Extraction Format)
  - LIB (pointers to the common libraries and common IP/G)

- Synthesis:

  - SYN (synthesis workplace)
  - TCL (synthesis scripts)
  - VER (formal verification, RTL vs. pre-layout workspace)
  - NET (synthesized or pre-layout and post-layout netlists)
  - MEM (memories timing and netlist files)
  - ENV (synthesis-related environmental files)
  - LIB (pointers to the common libraries and common IP/G)
  - RTL (design in RTL format)

- Physical:

  - TCL (place and route scripts)
  - NET (fully routed post-layout netlist)
  - PNR (place and route workspace)
  - VER (physical verification workplace)
  - MEM (memory compiler workspace)
  - SPEF (extraction files in Standard Parasitic Extraction Format)
  - GDS (ASIC design layout in GDS format)
  - ENV (physical design-related environmental files)
  - LEF (macros, standard cells abstract, and tech files in LEF format)
  - SDC (design constraint files)
  - CPF (power management file in Cadence Power Format)
  - DEF (netlist in Design Exchange Format)
  - QRC (extraction files)
  - NSE (noise analysis workspace)
  - LIB (pointers to the common and IP libraries)
  - PWR (power analysis workspace)
  - OA (open-access layout directory for physical design integration)

The example of *Perl* script, under the *Physical Design Scripts* section of this chapter named *make_dir.pl*, is used to create synthesis, timing, and physical design directories under the implementation directory. In addition, the script creates all their sub-directories.

The *make_dir.pl* script takes input command line:

```
make_dir.pl -p moonwalk
```

This command should be executed under the project directory (in this example, *moonwalk*); otherwise, it issues an ERROR.

In addition, the script copies a generic environmental file from the templates directory under common data structure in each of the ENV sub-directories of the project-specific directory. Here is an example of the *moonwalk* physical design named *moonwalk_pd_data.env*:

```
### moonwalk_pd_data.env ###
### Process Technology ###

Process node20um


###Candance Hspice Noise Analysis Paths ###
HSpiceLib \
/common/tools/external/cds/node20um/models/hspice/node20.lib
HSpiceRes \
/common/tools/external/cds/node20um/models/hspice/ResModel.spi
HSpiceMod \
/common/tools/external/cds/node20um/models/hspice/Hspice.mod

### Cadence OA Devices & Tech_file ###
CdsOADevices  /common/tools/external/cds/oa/node20um/devices
CdsOATechFile /common/tools/external/cds/oa/tech/node20.tf

### Common Libraries Directory Path ###
Release /common

### Worst-case and Best-case ASIC Libraries ###
SCLibMaxMin \
node20_fn_sp_125c_1p0v_ss:node20_fn_sp_m40c_1p5v_ff
IOLibMaxMin \
io35um_fn_sp_125c_1p0v_ss:io35um_fn_sp_m40c_1p5v_ff

### LEF Views Libraries Paths ###
SClefFile /node20um/stdcells/lef/node20_stdcells.lef
IOlefFile /node20um/pads/lef/io35um.lef

### Verilog Views ASIC Libraries Paths ###
SClibVerilogPath /node20um/stdcells/verilog/node20_stdcells.v

IOlibVerilogPath  /libraries/node20um/pads/verilog/io_35um.v


### CDL Views ASIC Libraries Paths ###
SClibCdlPath  /node20um/stdcells/cdl/node20_stdcells.cdl
SClibCdlPath  /node20um/pads/cdl/io_35um.cdl
```

```
### Cadence OA (layout) Views ASIC Libraries Paths ###
SClibCdsOAPath          /libraries/node20um/stdcells/oa/DFII/node20_
stdcells
IOlibCdsOAPath  /libraries/node20um/pads/oa/DFII/io_35um

### Power Management Hard and/or Soft  (true/false) ###
PowerManagement true

### Hierarchical Soft Macro Module Names ###
HieModule clk_gen

### Project Directory and Name ###
Path /project
Project moonwalk

### Netlist File and Top-Level Name ###
NetlistName moonwalk_pre_layout.vg
TopLevelName moonwalk

### MacroName:NumberOfBlockage:Gds/Lef View ###
Macros A2D:8:G PLL:8:G ram_288x72:4:L

### MinRouteLayer:MaxRouteLayer ###
MinMaxLayer 1:8

### X and Y die size in Micron ###
DieX 6080
DieY 6740


PadFile moonwlk_pad.def
```

This environmental template is specific to a physical design and structured so that the first column is the keyword and next column is data related to the project. It is used to create physical design *Tcl* scripts using *Perl* script (*make_pd_tcl.pl*).

The next step is to modify *moonwalk_pd_data.env* to project-specific requirements such as tools-related, libraries, project name, top-level design name, etc. Afterward, the user needs to execute the *make_pd_tcl.pl* (an example of this *Perl* program is under *Physical Design Scripts* section of this chapter) to create physical design scripts for generating related *Tcl* scripts such as floorplan, placement, Clock Tree Synthesis, routing, ECO, export post-layout netlist, extraction files, and layout file (GDS). The command used is:

```
make_pd_tcl.pl -input moonwalk_pd_data.env
```

This command creates all physical design *Tcl* scripts. These *Tcl* scripts are:

- moonwalk_cofig.tcl: Tools configuration settings
- moonwalk_setting.tcl: Design-related global settings
- moonwalk_view_def.tcl: MMMC view definition
- moonwalk_flp.tcl: Floorplan *Tcl*
- moonwalk _plc.tcl: Placement *Tcl*
- moonwalk _cts.tcl: Clock Tree Synthesis (CTS) *Tcl*
- moonwalk _frt.tcl: Final route *Tcl*
- moonwalk _mmc.tcl: MMMC timing violation fix *Tcl*
- moonwalk_eco.tcl: Engineering Order Change (ECO) *Tcl*
- moonwalk _net.tcl: Netlist export *Tcl*
- moonwalk _spef.tcl: Extraction (SPEF) export *Tcl*
- moonwalk _gds.tcl: Layout (GDS) export *Tcl*

1.4   Process Variation Impacts

Typical timing analysis consists of examining the operation of the design across a range of process, voltage, and temperature conditions.

In the past, it was acceptable to expect that the operational space of a design could be bounded by analyzing the design at two different points. The first point was chosen by taking the worst-case condition for all three operating condition parameters (process, voltage, temperature), and the second point was chosen by taking the best-case conditions for the same three parameters.

Operating condition domain for timing signoff was assumed that the worst slack for a design in our three-dimensional box was covered by just two points because delay was typically monotonic along the axis between the points. And since delay at earlier process nodes was dominated by cell delay, there was little thought to looking at the process range for interconnect layers; therefore, parasitic files were typically extracted at the nominal process.

When process nodes moved to lower geometries, like 65 and 45 nm, supply voltage levels were reduced to 1.0 V or less, and thus temperature began to have a contrarian effect on cell delay. Whereas in previous technology node delay increased with high temperature, temperature inversion effects began to increase delay at lower temperatures under worst-case voltage and process conditions.

The direct result for timing signoff was an increase in the number of corners for timing analysis and closure and optimization. The number of operating corners quickly doubled from two to four or even five for those who insisted on timing test modes at typical operating conditions.

Process variations in metallization and devices such N-channel and P-channel (PMOS and NMOS transistors) now had a non-negligible impact on the timing of the design. Plus, metal line width became small enough to impact the resistance of the wire

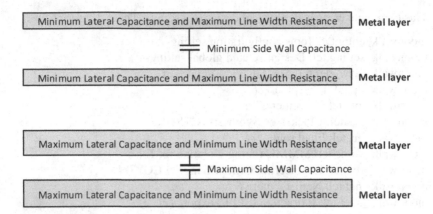

Fig. 1.3 Metal layer corners for capacitance and resistance

with just a small amount of variation. Given that metallization is a separate process from *base* layer (PMOS and NMOS transistors) processing, engineers could not assume that process variation tracked in the same direction for both base and metal layers.

For advanced technology at 40 nm and to a larger extent at 20 nm node and below, multiple extraction corners were now required for timing analysis and optimization. These extraction corners consisted of maximum capacitance and minimum resistance (Cmax and Rmin), minimum capacitance and maximum resistance (Cmin and Rmax), minimum capacitance and minimum resistance (Cmin and Rmin), maximum capacitance and maximum or resistance (Cmax and Rmin), and typical.

Cmax and Cmin are a summation of both their minimum and maximum lateral and sidewall capacitance as shown in Fig. 1.3.

The minimum (best) capacitance and minimum (best) resistance and the maximum (worst) capacitance and maximum (worst) resistance are considered artificial corners. For these two corners, according to the law of physics, minimum and maximum for both capacitance and resistance will not occur simultaneously. However, one could use these two artificial corners eliminating Cmax/Rmin and Cmin/Rmax during design timing closure.

In addition to metal layer process variations, base layers also would be impacted by process variation. For process node 20 nm and below, the base layer becomes more sensitive to process variation and temperature due to their small PMOS and NMOS transistor geometry which affects their transitional (i.e., high-to-low and low-to-high) threshold sensitivity. Figure 1.4 shows PMOS and NMOS transistor timing corners without their timing variations.

Summarizing the process variations at 20 nm and below, the following corners may be considered:

- PMOS/NMOS Transistor: fast-fast, fast-slow, slow-fast, slow-slow, and typical
- Interconnects: RminCmax, RmaxCmin, RCmax, RCmin, and typical
- Voltage: Vmin, Vmax
- Temperature: Tmax, Tmin

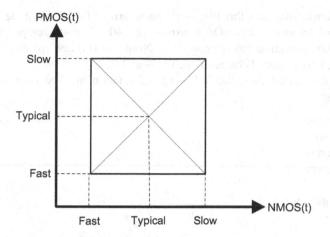

Fig. 1.4 PMOS and NMOS transistor timing corners

To calculate the number of combinations of these variations in the above process, voltage, and temperature (PVT) for timing analysis, it would be $2 \times 5 \times 5 \times 2 = 100$ corners for one mode (i.e., functional setup). However, this calculation must be done again for each additional mode in the design.

Having a consistent naming convention, whether it is directory name or data files, allows users to locate design directories and related data files and their contents with ease as well as enables them to develop productivity scripts that can be shared among users for different ASIC projects for rapid timing analysis and closure for all functions and corners.

Another benefit of having naming conventions, along with usage scripting for project data structures and related data, is that it creates a unified ASIC design and implementation methodology. Having an advanced unified ASIC design implementation requires not only programming but also discipline.

To avoid confusion, naming conventions are used to convey the contents of data files. The files can be libraries and other project-related data files.

For libraries, data files such as Liberty (commonly known as *lib* files), one may consider the following as part of their naming convention:

- Process node
- Transistor type: Combination of fast and slow for PMOS and NMOS
- Parasitic extraction type: Combination of minimum and maximum for capacitance and resistance
- Operating temperature: Minimum and maximum
- Operating voltage: Minimum and maximum

Some suggested naming conventions for the *moonwalk* example project:
Library name:

```
node20_sp_nf_m40_1p05.lib
```

   This example indicates this library is characterized for 20 nm node with slow
PMOS transistor and fast NMOS transistor at −40 °C (m40) temperature under
1.05 V (1p05) operation voltage condition. Note that (m) and (p) were chosen to
avoid using UNIX special character such as − and +.

   For project-related files, the following information may be used for naming
conventions:

- Project name
- Process node
- Extraction
- Temperature
- Voltage
- Prefix data type
- Function

   The following are some naming convention examples for the *moonwalk* project:

Liberty Timing file:

```
node20_sp_fn_m40c_1p05v.lib.
```

This example indicates that the 20 nm standard cells were characterized with slow
    PMOS transistor and fast NMOS transistor at -40C temperature with 1.05V
    voltage.
Extraction: `moonwalk_node20_rmin_cmax.spef`
This example indicates the data was extracted using RminCmax (i.e., minimum
    resistance and maximum capacitance) for the 20 nm process node in SPEF.
Timing:

```
moonwalk_node20_sp_fn_rmin_cmax_m40c_1p05v_hold.rpt
```

This naming example indicates the result of STA report file for 20 nm process node
    with RminCmax extraction at -40C temperature with 1.05V operating voltage
    with slow PMOS transistor and fast NMOS transistor as a result of hold
    analysis.
Simulation:

```
moonwalk_node20_sp_fn_rmin_cmax_m40c_1p05v.sdf
```

This naming example indicates the simulation delay file was compiled for 20 nm
    process nodes using RminCmax extraction at -40C temperature at 1.05V operat-
    ing voltage with slow PMOS transistor and fast NMOS transistor for functional
    gate-level simulation in Standard Delay Format (SDF).
Layout data file:

```
moonwalk_node20_R1.gds
```

This naming example indicates the layout data exported in Geometrical Data Set (GDS) format for R1 revision. Note that the only dependency GDS has is the process node and its revision.

1.5 Physical Design Scripts

A *Perl* script (*make_implementation_dir.pl*) that creates timing, synthesis, and physical directories under implementation directory is shown below. This example only generates a partial sub-directory according to Fig. 1.2 under physical directory:

```perl
#!/usr/local/bin/perl
use Time::Local;
system(clear);

### make_implemention_dir.pl ###


$pwd = `pwd`;
chop $pwd;

&parse_command_line;
@pwd = split(/\//,$pwd);
$i=@pwd-1;
unless($pwd[$i] =~ $ProjName)
   {die "ERROR: You must execute this program at $pwd/$ProjName.
\n"};

print(" "*** Setting Implementation Directory to $pwd/implementation
***\n";


$mode=0777;
$PhysicalDir = $pwd.'/implementation/physical';
$SynDir = $pwd.'/implementation/synthesis';
$TimDir = $pwd.'/implementation/timing';

@PD_SubDir = ('pnr','gds','net','spef','pver','pwr','mem','tcl','
pnr','env');


if(chdir($PhysicalDir)>0){print "\n"; die "ERROR: $PhysicalDir
exists \n";}
```

```perl
if(chdir($SynDir)>0){print "\n"; die "ERROR: $SynDir exists \n";}
if(chdir($TimDir)>0){print "\n"; die "ERROR: $TimDir exists \n";}

print("\n");


print("  Creating PHYSICAL directories...\n");
mkdir($PhysicalDir,$mode);
print("\n");

print("  Creating SYNTHESIS directories...\n");
mkdir($SynDir,$mode);
print("\n");

print("  Creating TIMING directories...\n");
mkdir($TimDir,$mode);
print("\n");

for($i=0; $i<@PD_SubDir; $i++){
   $dir = $pwd.'/implementation/physical'.$PD_SubDir[$i];
   mkdir($dir,$mode);}

$         T         r         a         g         e         t
= $pwd.'/implementation/physical/env/'.$Projectname.'_pd.env';
print("Coping Project Environmental Template to $Target\n");
open(Input,"/repository/templates/pd_env_dat.env";
open(output,">$Target");

unless ($ProjName)
     { print "ERROR: incorrectly specified command line. Use -h for
more information.\n";
         exit(0); } }

sub print_usage {
    print "\usage: create_impementaion_dir -p XX\n";
    print "            -p  # Where XX is Project name. \n";
    print "\n";
   print "create all directories of Physical Design: /project/XX/
physical/\n"}
```

Perl Script to Generate *Tcl* Scripts

make_pd_tcl.pl generates all *Tcl* scripts that are required for a physical design (i.e., floorplan, placement, Clock Tree Synthesis, and final route):

```perl
#!/usr/local/bin/perl
use Time::Local;
```

```perl
system(clear);

### make_pd_tc.pl ###


&parse_command_line;


unless (open(Input,$DataFile))
   { die "ERROR: could not open Input file : $DataFile\n";}

### setting default values ###
$PwrMgt = 0;
### CTS Buffers and CTS Inverters ###
$CTSBuf = "ckbufd24 ckbufd16 ckbufd12 ckbufd10";
@CTS_buf_lst = split(/ +/,$CTSBuf);
$CTSInv = "ckinv_lvtd24 ckinv_lvtd16 ckinv_lvtd12 ckinv_lvtd10";
@CTS_inv_lst = split(/ +/,$CTSInv);

### hold Buffers and Delay Cells ###
$HoldBuf = "bufd2 dly2d1";
@Hold_buf_lst = split(/ +/,$HoldBuf);

### Pad and StdCells Fillers ###
$line =  "pd_fil25 pd_fil20 pd_fil10";
@Padstd_fil = split(/ +/,$line);
$line = "std_fil64 std_fil32 std_fil16";
@Stdstd_fil = split(/ +/,$line);

### Do Not Use Library*/*CellType* ###
$line   =  "stdcells_*/*d0p5   stdcells_*/*d0   stdcells_*/*dly*
stdcells_lvt*/*";
@DontUse = split(/ +/,$line);

### Spare Cell Types ###
$line = "sdfsrd4 nd2d4 mux2d4 aoi22d4 bufd24 invd12 ";
@SpareCells = split(/ +/,$line);

#### Add Instance Name ###
$AddInst[0] = "LOGO";

### Skip Route NETS (Hand Routed) ###
$line = "PLL_VVD PLL_VSS IO_VDDO IO_VSSO";
@SkipNet = split(/ +/,$line);
```

```
### Do Not Touch Cells ###
$line = "*spare* *stdcells_dt* stdcells_dt*";
@KeepCells  = split(/ +/,$line)

### MinRouteLayer:MaxRouteLayer ###
$line = "1:8";
split(/:/,$line);
$MinLayer = $_[0];
$MaxLayer = $_[1];
### From form Physical Design Environmental File ###
do {
    $line = <Input> ;
    chop($line) ;
    while($line =~ m/\\$/) {
        $line =~ s/\\$//;
        $line = $line.<Input>;
        chop($line);
    }
    $line =~ s/\#.*//;
    $line =~ s/^\s+//;
    $line =~ s/\s+$//;
    $line =~ s/\s+/ /g;
    if($line ne "") { print "$line\n"; }
    @line = split(/ +/,$line);

    if ($line[0] =~ /^SClefFile\b/) {
      @line = split(/\//,$line[1]);
      for($i=0;$i<@line;$i++) {
        $SCLibName = $line[$i];
      }
        push(@SCLib,$SCLibName);
        push(@RefLib,$SCLibName);
    }

    if ($line[0] =~ /^IOlefFile\b/) {
      @line = split(/\//,$line[1]);
      for($i=0;$i<@line;$i++) {
        $IOLibName = $line[$i];
      }
        push(@IOLib,$IOLibName);
        push(@RefLib,$IOLibName);
    }

    if ($line[0] =~ /^SCLibMaxMin\b/) {
      for($i=1;$i<@line;$i++) {
```

```
            split(/:/,$line[$i]);
            $SClibWCDB = $_[0];
            $SClibBCDB = $_[1];
            push(@SCMaxLib,$SClibWCDB);
            push(@SCMinLib,$SClibBCDB);
        }
    }

    if ($line[0] =~ /^IOLibMaxMin\b/) {
      for($i=1;$i<@line;$i++) {
        split(/:/,$line[$i]);
        $IOlibWCDB = $_[0];
        $IOlibBCDB = $_[1];
        push(@IOMaxLib,$IOlibWCDB);
        push(@IOMinLib,$IOlibBCDB);
      }
    }
    if ($line[0] =~ /^HieModule\b/) {
      for($i=1;$i<@line;$i++) {
        $HieModule[$i] = $line[$i];}
    }

    if ($line[0] =~ /^Path/){$Path = $line[1]}
    if ($line[0] =~ /^PnrDir/){$PnrDir = $line[1]}
    if ($line[0] =~ /^Process/){$Process = $line[1]}
    if ($line[0] =~ /^Library/){$Library = $line[1]}
    if ($line[0] =~ /^Project/){$Project = $line[1]}
    if ($line[0] =~ /^NetlistName/){$NetlistName = $line[1]}
    if ($line[0] =~ /^TopLevelName/){$TopLevelName = $line[1]}
    if ($line[0] =~ /^DieX/){$DieX = $line[1]}
    if ($line[0] =~ /^DieY/){$DieY = $line[1]}
    if ($line[0] =~ /^Macros/){ for ($i=1; $i<@line; $i++)
                                  { push(@Macros,$line[$i])}}
    if ($line[0] =~ /^PowerManagement/)
        {if($line[1] =~ /^true/){$PwrMgt = 1}}
  } until (eof(Input));

### QRC Extraction Files and Cap Tables ###


$CapTableWorst  = "$Path/tools/external/cds/qrc/worst/${Process}.
capTbl";
$CapTableBest   =  "$Path/tools/external/cds/qrc/best/${Process}.
capTbl";
$QRCWorst = "$Path/tools/external/cds/qrc/worst/qrcTechFile";
```

```perl
$QRCBest = "$Path/tools/external/cds/qrc/best/qrcTechFile";

$WC_condition = $SCMaxLib[0];
$BC_condition = $SCMinLib[0];

$Power = "VDD";
$Ground = "VSS";

push(@RefLib,"MACROS");
print"\n";
if (@RefLib>0 ) { print "RefLib List: "}
for ($i=0; $i<@RefLib; $i++){print "$RefLib[$i] "}
print"\n";

print"\n";
if (@HieModule>0 ) { print "HieModule List: "}
for ($i=1; $i<@HieModule; $i++){print "$HieModule[$i] "}
print"\n";

print"\n";
if (@Macros>0 ) { print "Macros List: "}
for ($i=0; $i<@Macros; $i++){
   @_ = split(/:/,$Macros[$i]);
   push(@MacroBlkg,$_[1]);
   push(@MacroGdsOrLef,$_[2]);
   $Macros[$i] =~ s/:.*//;
   print "$Macros[$i]\n "}
print"\n";

if (@AddInst>0 ) { print "Instance List:\n"}
for ($i=0; $i<@AddInst; $i++){
   @_ = split(/:/,$AddInst[$i]);
   push(@InstName,$_[0]);
   push(@InstType,$_[1]);
   push(@InstLib,$_[2]);
      print "$InstName[$i]  of  Type  $InstType[$i]  and  library
$InstLib[$i] \n"}

print"\n";
if (@SkipNet>0 ) { print "SkipNet List: "}
for ($i=0; $i<@SkipNet; $i++){print "$SkipNet[$i] "}
print"\n";

if ($PadFile eq ""){ ($Ldb = $TopLevelName ) =~ tr/a-z/A-Z/}
else {$Ldb = 'MOONWALK'}
```

```
close(Input);


### Design (MMMC) view_definition.tcl ###
$Output      =      $Path.'/'.$Project.'/implemention/physical/
TCL/'.${TopLevelName}.'_view_def.tcl';
if(!(-e $Output)) {
  print "Building view_definition.tcl --> $Output\n\n";
unless (open(ViewDefFile,">$Output")) {
    die "ERROR: could not create view_definition file : $Output\n";}

  print(ViewDefFile "###\n");
   print(ViewDefFile "Puts \"... Multi-Mode Multi-Corner Timing
...\"\n\n");
  print(ViewDefFile "create_rc_corner -name rc_max \\\n");
  print(ViewDefFile "   -T 125 \\\n");
  print(ViewDefFile "   -qx_tech_file $QRCWorst \\\n");
  print(ViewDefFile "   -preRoute_res 1.00 \\\n");
  print(ViewDefFile "   -preRoute_cap 1.00 \\\n");
  print(ViewDefFile "   -postRoute_res 1.00 \\\n");
  print(ViewDefFile "   -postRoute_cap 1.00 \\\n");
  print(ViewDefFile "   -postRoute_clkres 1.00 \\\n");
  print(ViewDefFile "   -postRoute_clkcap 1.00 \\\n");
  print(ViewDefFile "   -postRoute_xcap 1.00 \n\n");

  print(ViewDefFile "create_rc_corner -name rc_min \\\n");
  print(ViewDefFile "   -T -40 \\\n");
  print(ViewDefFile "   -qx_tech_file $QRCBest \\\n");
  print(ViewDefFile "   -preRoute_res 1.00 \\\n");
  print(ViewDefFile "   -preRoute_cap 1.00 \\\n");
  print(ViewDefFile "   -postRoute_res 1.00 \\\n");
  print(ViewDefFile "   -postRoute_cap 1.00 \\\n");
  print(ViewDefFile "   -postRoute_clkres 1.00 \\\n");
  print(ViewDefFile "   -postRoute_clkcap 1.00 \\\n");
  print(ViewDefFile "   -postRoute_xcap 1.00 \n\n");

   print(ViewDefFile "set libpath ${Path}/${Project}/physical/lib
\n\n");


  print(ViewDefFile "create_library_set -name ss_libs \\\n");
  print(ViewDefFile "   -timing [list \\\n");
  for ($i=0;$i<@SCMaxLib;$i++){
     print(ViewDefFile " \${libpath}\/$SCMaxLib[$i].lib \\\n")}
  for ($i=0;$i<@IOMaxLib;$i++){
```

```
    print(ViewDefFile " \${libpath}\/$IOMaxLib[$i].lib \\\n")}
  for ($i=0; $i<@Macros; $i++){
    print(ViewDefFile " \${libpath}\/$Macros[$i]_ss.lib \\\n")}
  for ($i=1; $i<@HieModule; $i++){
    print(ViewDefFile " \${libpath}\/$HieModule[$i]_ss.lib \\\n")}
  print(ViewDefFile "              ] \n\n");

  print(ViewDefFile "create_library_set -name ff_libs \\\n");
  print(ViewDefFile "   -timing [list \\\n");
  for ($i=0;$i<@SCMinLib;$i++){
    print(ViewDefFile " \${libpath}\/$SCMinLib[$i].lib \\\n")}
  for ($i=0;$i<@IOMinLib;$i++){
    print(ViewDefFile " \${libpath}\/$IOMinLib[$i].lib \\\n")}
  for ($i=0; $i<@Macros; $i++){
    print(ViewDefFile " \${libpath}\/$Macros[$i]_ff.lib \\\n")}
  for ($i=1; $i<@HieModule; $i++){
    print(ViewDefFile " \${libpath}\/$HieModule[$i]_ff.lib \\\n")}
  print(ViewDefFile "              ] \n\n");

  print(ViewDefFile "create_delay_corner \\\n");
   print(ViewDefFile "  -name slow_max -library_set ss_libs -rc_
corner rc_max\n");
  print(ViewDefFile "create_delay_corner \\\n");
  print(ViewDefFile " -name fast_min -library_set ff_libs -rc_cor-
ner rc_min\n\n");

  print(ViewDefFile "set active_corners [all_delay_corners]\n");
   print(ViewDefFile "if {[lsearch \$active_corners slow_max] !=
-1} { \n");
  print(ViewDefFile " set_timing_derate \\\n");
   print(ViewDefFile "   -data -cell_delay -early -delay_corner
slow_max 0.97\n");
  print(ViewDefFile " set_timing_derate \\\n");
   print(ViewDefFile "   -clock -cell_delay -early -delay_corner
slow_max 0.97\n");
  print(ViewDefFile " set_timing_derate \\\n");
  print(ViewDefFile " -data -cell_delay -late -delay_corner slow_
max 1.03 \n");
  print(ViewDefFile " set_timing_derate \\\n");
  print(ViewDefFile " -clock -cell_delay -late -delay_corner slow_
max 1.03\n");
  print(ViewDefFile " set_timing_derate \\\n");
  print(ViewDefFile " -data -net_delay -early -delay_corner slow_
max 0.97\n");
  print(ViewDefFile " set_timing_derate \\\n");
```

```
  print(ViewDefFile " -clock -net_delay -early -delay_corner slow_
max 0.97\n");
  print(ViewDefFile " set_timing_derate \\\n");
  print(ViewDefFile " -data -net_delay -late -delay_corner slow_
max 1.03\n");
  print(ViewDefFile " set_timing_derate \\\n");
  print(ViewDefFile " -clock -net_delay -late -delay_corner slow_
max 1.03\n");
  print(ViewDefFile "} \n\n");


  print(ViewDefFile "if {[lsearch \$active_corners fast_min] !=
-1} { \n");
  print(ViewDefFile " set_timing_derate \\\n");
  print(ViewDefFile " -data -cell_delay -early -delay_corner fast_
min 0.95\n");
  print(ViewDefFile " set_timing_derate \\\n");
  print(ViewDefFile " -clock -cell_delay -early -delay_corner
fast_min 0.97\n");
  print(ViewDefFile " set_timing_derate \\\n);
  print(ViewDefFile " -data -cell_delay -late -delay_corner fast_
min 1.05\n");
  print(ViewDefFile " set_timing_derate \\\n");
  print(ViewDefFile " -clock -cell_delay -late -delay_corner fast_
min 1.05\n");
  print(ViewDefFile " set_timing_derate \\\n");
  print(ViewDefFile " -data -net_delay -early -delay_corner fast_
min 0.97\n");
  print(ViewDefFile " set_timing_derate \\\n");
  print(ViewDefFile " -clock -net_delay -early -delay_corner fast_
min 0.97\n");
  print(ViewDefFile " set_timing_derate \\\n");
  print(ViewDefFile " -data -net_delay -late -delay_corner fast_
min 1.05\n");
  print(ViewDefFile " set_timing_derate \\\n");
  print(ViewDefFile " -clock -net_delay -late -delay_corner fast_
min 1.05\n");
  print(ViewDefFile "} \n\n");


  print(ViewDefFile "create_constraint_mode -name setup_func_mode
\\\n");
  print(ViewDefFile "    -sdc_files \\\n";
  print(ViewDefFile "[list ${Path}/${Project}/physical/SDC/\\\n");
  print(ViewDefFile "${TopLevelName}_func.sdc] \n\n");


  print(ViewDefFile "create_constraint_mode -name hold_func_mode
```

```
\\\n");
    print(ViewDefFile "      -sdc_files [list ${Path}/${Project}/
physical/SDC/\\\n");
  print(ViewDefFile "${TopLevelName}_func.sdc] \n\n");


print(ViewDefFile   "create_constraint_mode  -name   setup_func_
mode\\\n");
    print(ViewDefFile "      -sdc_files [list ${Path}/${Project}/
physical/SDC/\\\n");
  print(ViewDefFile "${TopLevelName}_func.sdc] \n\n");


print(ViewDefFile   "create_constraint_mode   -name    hold_func_
mode\\\n");
    print(ViewDefFile "      -sdc_files [list ${Path}/${Project}/
physical/SDC/\\\n");
  print(ViewDefFile "${TopLevelName}_func.sdc] \n\n");

  print(ViewDefFile "create_constraint_mode -name hold_scanc_mode
\\\n");
    print(ViewDefFile "      -sdc_files [list ${Path}/${Project}/
physical/SDC/\\\n);
  print(ViewDefFile "${TopLevelName}_scanc.sdc] \n\n");

  print(ViewDefFile "create_constraint_mode -name hold_scans_mode
\\\n");
    print(ViewDefFile "      -sdc_files [list ${Path}/${Project}/
physical/SDC/\\\n");
  print(ViewDefFile "${TopLevelName}_scans.sdc] \n\n");

  print(ViewDefFile "create_constraint_mode -name setup_mbist_mode
\\\n");
    print(ViewDefFile "      -sdc_files [list ${Path}/${Project}/
physical/SDC/\\\n");
  print(ViewDefFile "${TopLevelName}_mbist.sdc] \n\n");

  print(ViewDefFile "create_constraint_mode -name hold_mbist_mode
\\\n");
    print(ViewDefFile "      -sdc_files [list ${Path}/${Project}/
physical/SDC/\\\n");
  print(ViewDefFile "${TopLevelName}_mbist.sdc] \n\n");

  print(ViewDefFile "if {[lsearch [all_analysis_views] hold_func]
== -1} { \n");
```

```
    print(ViewDefFile "       create_analysis_view -name hold_func
\\\n");
  print(ViewDefFile "       -constraint_mode hold_func_mode \\\n");
  print(ViewDefFile "       -delay_corner fast_min \\\n}\n");

  print(ViewDefFile "if {[lsearch [all_analysis_views] setup_func]
== -1} { \n");
    print(ViewDefFile "     create_analysis_view -name setup_func
\\\n");
  print(ViewDefFile "       -constraint_mode setup_func_mode \\\n");
  print(ViewDefFile "       -delay_corner slow_max \\\n}\n");

  print(ViewDefFile "if {[lsearch [all_analysis_views] hold_func]
== -1} { \n");
    print(ViewDefFile "     create_analysis_view -name hold_func
\\\n");
  print(ViewDefFile "       -constraint_mode hold_func_mode \\\n");
  print(ViewDefFile "       -delay_corner fast_min \\\n}\n");

  print(ViewDefFile "if {[lsearch [all_analysis_views] setup_func]
== -1} { \n");
    print(ViewDefFile "     create_analysis_view -name setup_func
\\\n");
  print(ViewDefFile "       -constraint_mode setup_func_mode \\\n");
  print(ViewDefFile "       -delay_corner slow_max \\\n}\n");

  print(ViewDefFile "if {[lsearch [all_analysis_views] hold_scanc]
== -1} { \n");
    print(ViewDefFile "     create_analysis_view -name hold_scanc
\\\n");
  print(ViewDefFile "       -constraint_mode hold_scanc_mode \\\n");
  print(ViewDefFile "       -delay_corner fast_min \\\n}\n");

  print(ViewDefFile "if {[lsearch [all_analysis_views] hold_scans]
== -1} { \n");
    print(ViewDefFile "     create_analysis_view -name hold_scans
\\\n");
  print(ViewDefFile "       -constraint_mode hold_scans_mode \\\n");
  print(ViewDefFile "       -delay_corner fast_min \\\n}\n");
    print(ViewDefFile "if {[lsearch [all_analysis_views] setup_
mbist] == -1} { \n");
    print(ViewDefFile "     create_analysis_view -name setup_mbist
\\\n");
    print(ViewDefFile "          -constraint_mode setup_mbist_mode
\\\n");
```

```
  print(ViewDefFile "        -delay_corner slow_max \\\n}\n");

  print(ViewDefFile "if {[lsearch [all_analysis_views] hold_mbist]
== -1} { \n");
   print(ViewDefFile "      create_analysis_view -name  hold_mbist
\\\n");
  print(ViewDefFile "      -constraint_mode hold_mbist_mode \\\n");
  print(ViewDefFile "        -delay_corner fast_min \\\n}\n");

  print(ViewDefFile "set_analysis_view \\\n");
  print(ViewDefFile " -setup [list setup_func] -hold [list hold_
func] \n");

  close(ViewDefFile);
}
else {print "Not building view_definition.tcl file --> $Output File
exists\n\n";}


### Design setting.tcl ###
$Output        =        $Path.'/'.$Project.'/implementaion/physical/
TCL/'.${TopLevelName}.'_setting.tcl';
if(!(-e $Output)) {
  print "Building setting file --> $Output\n\n";
  unless (open(SetFile,">$Output"))
    {die "ERROR: $Output File exists\n";}

  print(SetFile "### Physical Design Setting ###\n");
  print(SetFile "setDesignMode -process 180\n\n");
  print(SetFile "#set_option liberty_always_use_nldm true\n\n");
  print(SetFile "set_interactive_constraint_modes \\\n);
  print(SetFile "[all_constraint_modes -active]\n\n");
  print(SetFile "#### Keep Instances and Do Not Use cells ####\n");
  if (@KeepCells >0) {
     for ($i=0; $i<@KeepCells; $i++){
    print(SetFile "set_dont_touch [get_cells -hier $KeepCells[$i]]
true\n");
     }
  }
  print(SetFile "\nforeach cell [list \\\n");
  for ($i=0;$i<@DontUse;$i++){
     print(SetFile "                    {$DontUse[$i]} \\\n");}
  print(SetFile "                ] { \n");
  print(SetFile "  setDontUse \$cell true \n");
  print(SetFile "} \n\n");
```

```
  print(SetFile "foreach net [list \\\n");
  print(SetFile "                      VDDO \\\n");
  print(SetFile "                   ] {\n");
  print(SetFile "   set_dont_touch [get_nets \$net] true\n");
  print(SetFile "   setAttribute -net \$net -skip_routing true\n");
  print(SetFile "}\n");
  close(SetFile);
}
else {print "Not building setting.tcl file --> $Output File
exists\n\n";}


### Design ignore_pins.tcl ###
$Output        =        $Path.'/'.$Project.'/implementation/physical/
TCL/'.${TopLevelName}.'_ignore_pins.tcl';
if(!(-e $Output)) {
  print "Building setting file --> $Output\n\n";
  unless (open(IgnorSetFile,">$Output"))
    {die "ERROR: $Output File exists\n";}

  print(IgnorSetFile "#set_ccopt_property \\\n");
   print(IgnorSetFile "sink_type ignore -pin dig_top/pll_reg_0/
CK\n");
  close(IgnorSetFile);
}
else {print "Not building ignore_pins.tcl file --> $Output File
exists\n\n";}


### Design reset_ignore_pins.tcl ###
$Output        =        $Path.'/'.$Project.'/implementation/physical/
TCL/'\\\n");
print(".${TopLevelName}.'_reset_ignore_pins.tcl';
if(!(-e $Output)) {
  print "Building setting file --> $Output\n\n";
  unless (open(ResetIgnorFile,">$Output")) {
    die "ERROR: could not create setting file : $Output\n";}

  print(ResetIgnorFile "#set_ccopt_property \\\n");
  print(" sink_type auto -pin dig_top/pll_reg_0/CK\n");
  close(ResetIgnorFile);
}
else {print "Not building reset_ignore_pins.tcl file --> $Output
File exists\n\n";}


### clk_gate_disable.tcl ###
```

```
$Output        =        $Path.'/'.$Project.'/implementation/physical/
TCL/'.\\\n");
print("${TopLevelName}.'_clk_gate_disable.tcl';
if(!(-e $Output)) {
  print "Building setting file --> $Output\n\n";
  unless (open(DisableClkGateFile,">$Output")) {
      die "ERROR: could not create disable clock gating file :
$Output\n";}

  print(DisableClkGateFile "#set_disable_clock_gating_check dig_
top/g120/B1\n");
  close(DisableClkGateFile);
}
else {print "Not building clk_gate_disable.tcl file --> $Output
File exists\n\n";}

### Design config.tcl ###
$Output        =        $Path.'/'.$Project.'/implementation/physical/
TCL/.'\\\n");
 print(ConfigFile "${TopLevelName}.'_config.tcl';
if(!(-e $Output)) {
  print "Building config tcl file --> $Output\n\n";
  unless (open(ConfigFile,">$Output"))
     {die "ERROR: could not create config tcl file : $Output\n";}

  print(ConfigFile "setMultiCpuUsage -localCpu 8\n\n");


 print(ConfigFile "#Attribute 'max_fanout' Not Found in library\n");
  print(ConfigFile "suppressMessage \"TECHLIB-436\"\n\n");

  print(ConfigFile "set_global report_timing_format \\\n");
   print(ConfigFile "{instance cell pin fanout load delay arrival
required}\n\n");

    print(ConfigFile "#Disables loading ECSM data from timing
libraries\n");
    print(ConfigFile "set_global timing_read_library_without_ecsm
true\n\n");
  print(ConfigFile "#set delaycal_use_default_delay_limit 1000\n");
  print(ConfigFile "\n");

  print(ConfigFile "#temporary increase\n");
  print(ConfigFile "setMessageLimit 1000 ENCDB 2078\n");
  print(ConfigFile "\n");
```

```
 print(ConfigFile "###\n");

  for ($i=1; $i<@HieModule; $i++){
     print(ConfigFile "#set rda_Input(ui_ilmdir)\\\n");
        print(ConfigFile "  \"$HieModule[$i]\"  ./$HieModule[$i].
ilm\n");}
  print(ConfigFile "\n");

  print(ConfigFile "alias h history\n");
  print(ConfigFile "\n");

  print(ConfigFile "### Data Base Process Modules ###\n");
  print(ConfigFile "\n");

    close(ConfigFile);
}
else {print "Not building config file --> $Output File exists\n\n";}

### Design Floorplan flp.tcl ###
$Output      =      $Path.'/'.$Project.'/implementation/physical/
TCL/'.$TopLevelName.'_flp.tcl';
if(!(-e $Output)) {
  print "Building FLP tcl file --> $Output\n\n";
  unless (open(FlpFile,">$Output")) {
    die "ERROR: could not create FLP tcl file : $Output\n";}

  print(FlpFile "### Floorplan Setup Environment ###\n");
  print(FlpFile "\n");
  print(FlpFile "source $Path/$Project/physical/TCL/\\\n");
  print(FlpFile "${TopLevelName}_config.tcl\n");
  print(FlpFile "\n");
  print(FlpFile "\n");

  print(FlpFile "foreach $dir \\\n");
  print(FlpFile "[list \\\n");
  print(FlpFile " ../rpt ../RPT/plc ../RPT/mmc ../$Ldb ../logs]
{\n");
  print(FlpFile " if { ! [file isdirectory \$dir] } {\n");
  print(FlpFile "   exec mkdir \$dir\n");
  print(FlpFile " }\n");
  print(FlpFile "}\n\n");

  print(FlpFile "### Initialize Design ###\n");
  print(FlpFile "\n");
  print(FlpFile "set init_layout_view \"\"\n");
```

```
print(FlpFile "set init_oa_view \"\"\n");
print(FlpFile "set init_oa_lib \"\"\n");
print(FlpFile "set init_abstract_view \"\"\n");
print(FlpFile "set init_oa_cell \"\"\n\n");

print(FlpFile "set init_gnd_net {VSS VSSO}\n");
print(FlpFile "set init_pwr_net {VDD VDDO}\n");
print(FlpFile "\n");

print(FlpFile "set init_lef_file \" \\\n");
print(FlpFile "  ${Path}/${Project}/physical/LEF/node20_tech.lef
\\\n");
print(FlpFile "   ${Path}/${Project}/physical/lef/clock_NDR.lef
\\\n");
for ($i=0;$i<@SCLib;$i++) {
     print(FlpFile " ${Path}/${Project}/physical/lef/$SCLib[$i]
\\\n");}
for ($i=0;$i<@IOLib;$i++) {
     print(FlpFile " ${Path}/${Project}/physical/lef/$IOLib[$i]
\\\n");}
for ($i=0;$i<@Macros;$i++) {
    print(FlpFile " ${Path}/${Project}/physical/lef/$Macros[$i].
lef \\\n");}
for ($i=1;$i<@HieModule;$i++) {
   print(FlpFile " ${Path}/${Project}/physical/lef/$HieModule[$i].
lef \\\n");}
for ($i=0;$i<@AddInst;$i++) {
    print(FlpFile " ${Path}/${Project}/physical/lef/$AddInst[$i].
lef \\\n");}
print(FlpFile "\"\n\n");

print(FlpFile "set init_assign_buffer \"1\"\n\n");


print(FlpFile "set init_mmmc_file \"${Path}/${Project}/physical/
TCL/\\\n");
print(FlpFile "${TopLevelName}_view_def.tcl\"\n");
print(FlpFile "set init_top_cell $TopLevelName \n");
  print(FlpFile "set init_verilog ${Path}/${Project}/physical/
net/\\\n");
print(FlpFile "$NetlistName \n\n");

print(FlpFile "init_design\n\n");
```

```
if ($PwrMgt) {
  print(FlpFile "#Load and commit CPF\n");
  print(FlpFile "loadCPF ${Path}/${Project}/physical/cpf/\\\n");
  print(FlpFile "${TopLevelName}.cpf\n");
  print(FlpFile "commitCPF\n\n");
}


  print(FlpFile "#Uncomment to preserve ports on any specific
module(s)\n");
  print(FlpFile "#set modules [get_cells -filter \\\n");
     print(FlpFile "\"is_hierarchical == true\" dig_top/
clk_gen*]\n");
  print(FlpFile "#getReport {query_objects \\\n");
  print(FlpFile "$modules -limit 10000} > ../keep_ports.list\n");
  print(FlpFile "#setOptMode -keepPort ../keep_ports.list\n\n");

  print(FlpFile "#Uncomment for N2N optimization\n");
  print(FlpFile "#source ${Path}/${Project}/physical/TCL/\\\n");
  print(FlpFile "${TopLevelName}_setting.tcl\n");
  print(FlpFile "#set_analysis_view \\\n");
  print(FlpFile "-setup [list func_setup] -hold [list func_hold
\n\n");

  print(FlpFile "#runN2NOpt -cwlm \\\n");
  print(FlpFile "#         -cwlmLib ../wire_load/mycwlm.flat \\\n");
  print(FlpFile "#          -cwlmSdc ../wire_load/mycwlm.flat.sdc
\\\n");
  print(FlpFile "#          -effort high \\\n");
  print(FlpFile "#          -preserveHierPinsWithSDC \\\n");
  print(FlpFile "#          -inDir n2n.input \\\n");
  print(FlpFile "#          -outDir n2n.output \\\n");
  print(FlpFile "#          -saveToDesignName n2n_opt.enc\n\n");

  print(FlpFile "#freedesign \n");
  print(FlpFile "#source ${Path}/${Project}/physical/TCL/\\\n");
  print(FlpFile "${TopLevelName}_config.tcl\n");
  print(FlpFile "#source n2n.enc\n");
  print(FlpFile "#source ${Path}/${Project}/physical/TCL/\\\n");
  print(FlpFile "${TopLevelName}_setting.tcl\n\n");

  print(FlpFile "#End of N2N optimization\n\n");


  print(FlpFile "deleteTieHiLo -cell TIELO\n");
  print(FlpFile "deleteTieHiLo -cell TIEHI\n\n");
```

```
print(FlpFile "source $Path/$Project/physical/TCL/\\\n");
print(FlpFile "${TopLevelName}_setting.tcl\n");
print(FlpFile "\n");

if ($PwrMgt == 0){
  print(FlpFile "globalNetConnect VDD -type pgpin -pin VDD\n");
  print(FlpFile "globalNetConnect VSS -type pgpin -pin VSS\n");
  print(FlpFile "globalNetConnect VSS -type pgpin -pin DVSS\n");
  print(FlpFile "\n");

  print(FlpFile "globalNetConnect VDD -type TIEHI\n");
  print(FlpFile "globalNetConnect VSS -type TIELO\n");
  print(FlpFile "\n");
}

print(FlpFile "saveDesign ../$Ldb/int.enc -compress\n");
print(FlpFile "\n");

print(FlpFile "### setup FloorPlan ###\n");
print(FlpFile "\n");

print(FlpFile "floorPlan \\\n");
print(FlpFile "  -siteOnly unit \\\n");
print(FlpFile "  -coreMarginsBy io \\\n");
print(FlpFile "  -d $DieX $DieY 12 12 12 12 \n");
print(FlpFile "\n");

if ($PadFile eq "") {}
else {
 print(FlpFile "addInst -cell $AddInst[0] -inst $AddInst[0]\n");
  print(FlpFile "\n");}

if ($PadFile eq "") {}
else {
  print(FlpFile "defIn $Path/$Project/physical/def/\\\n");
  print(ViewDefFile "${TopLevelName}_pad.def\n\n");
    print(FlpFile "#foreach side [list top left right bottom]
{\n");
  print(FlpFile "#  addIostd_filer \\\n");
  print(FlpFile "#     -prefix pd_filer \\\n");
  print(FlpFile "#     -side \$side \\\n");
  print(FlpFile "#     -cell {pd_fil10 pd_fil1 pd_fil01 pd_fil001}
\n");
  print(FlpFile "#}\n\n");}

if ($PadFile eq "") {
```

```
print(FlpFile "defIn $Path/$Project/physical/def/\\\n");
print(ViewDefFile "${TopLevelName}_pin.def\n\n");}

if ($PwrMgt) {
print(FlpFile "### Create SOFT Power Domain with Keep Out Area
###\n");
  print(FlpFile "setObjFPlanBox Group PDWN 988 468 1220 560\n");
   print(FlpFile "modifyPowerDomainAttr PDWN -minGaps 28 28 28
28\n");
    print(FlpFile "modifyPowerDomainAttr PDWN -rsExts 30 30 30
30\n\n");

  print(FlpFile "###  Create Nested Power Domains ###\n");
    print(FlpFile "#setObjFPlanBox Group PDWN2 1028 500 1130
522\n");
  print(FlpFile "#modifyPowerDomainAttr PDWN2 -minGaps 28 28 28
28\n");
   print(FlpFile "#modifyPowerDomainAttr PDWN2 -rsExts 30 30 30
30\n");
  print(FlpFile "#cutPowerDomainByOverlaps PDWN\n\n");

  print(FlpFile "### Create HARD Power Domain ###\n");
  print(FlpFile "\n");

  print(FlpFile "### Add Power Switch Ring ###\n");
  print(FlpFile "addPowerSwitch  ring \\\n");
  print(FlpFile "     -powerDomain PDWN \\\n");
   print(FlpFile "      -enablePinIn {PWRON1 } -enablePinOut
{PWRONACK1 } \\\n");
    print(FlpFile "        -enableNetIn {PWRDWN} -enableNetOut
{swack_1 } \\\n");
  print(FlpFile "     -specifySides {1 1 1 1} \\\n");
  print(FlpFile "     -sideOffsetList {3 3 3 3} \\\n");
  print(FlpFile "   -globalSwitchCellName {{thdrrng S} {hdrrng_
clamp D}} \\\n");
  print(FlpFile "     -bottomOrientation MY \\\n");
  print(FlpFile "     -leftOrientation MX90 \\\n");
  print(FlpFile "     -topOrientation MX \\\n");
  print(FlpFile "     -rightOrientation MY90 \\\n");
  print(FlpFile "     -cornerCellList hdrcor_outer \\\n");
   print(FlpFile "     -cornerOrientationList {MX90 MX MY90 MY}
\\\n");
    print(FlpFile "      -globalstd_filerCellName {{hdrrng_fill F}}
\\\n");
   print(FlpFile "     -insideCornerCellList thdrcor_inner \\\n");
```

```
    print(FlpFile "      -instancePrefix SWITCH_ \\\n");
    print(FlpFile "      -globalPattern {D S S S S S S S S S S D}
\n\n");

    print(FlpFile "set PDWN_SWITCH [addPowerSwitch \\\n");
    print(FlpFile "     -ring -powerDomain PDWN \\\n");
    print(FlpFile "     -getSwitchInstances]\n\n");
    print(FlpFile "rechainPowerSwitch \\\n");
    print(FlpFile "     -enablePinIn {PWRON2} \\\n");
    print(FlpFile "     -enablePinOut {PWRONACK2} \\\n");
    print(FlpFile "     -enableNetIn {swack_1} \\\n");
          -enableNetOut {swack_2} \\\n");
    print(FlpFile "     -chainByInstances \\\n");
    print(FlpFile "     -switchInstances \$PDWN_SWITCH\n\n");
  }

  print(FlpFile "setInstancePlacementStatus\\\n");
  print(FlpFile " -allHardMacros -status fixed\n\n");

  print(FlpFile "### CreateRegion ###\n");
  print(FlpFile "#createInstGroup clk_gen -isPhyHier\n");
      print(FlpFile   "#addInstToInstGroup   clk_gen   dig_top/
clk_gen/*\n");
  print(FlpFile "#createRegion clk_gen 1690 3810 2500 4080\n\n");

  print(FlpFile "### For Magnet Placement ###\n");
  print(FlpFile "#place_connected \\\n");
  print(FlpFile "   -attractor /dig_top/PLL \\\n");
  print(FlpFile "   -attractor_pin clk -level 1 -placed \n\n");

  print(FlpFile "#source $Path/$Project/physical/TCL/\\\n");
  print(FlpFile "${TopLevelName}_png.tcl\n\n");

  print(FlpFile "#Check for missing vias\n");
  print(FlpFile "verifyPowerVia -layerRange {MET4 MET1}\n\n");

    print(FlpFile   "defOut   -floorplan   $Path/$Project/physical/
def/\\\n");
  print(FlpFile "${TopLevelName}_flp.def\n");
  print(FlpFile "\n");

  Print(FlpFile "### Reuse the Floorplan with New Netlist ###\n");
  print(FlpFile "#defIn $Path/$Project/physical/def/\\\n");
  print(FlpFile "${TopLevelName}_flp.def\n\n");
```

```
  print(FlpFile "timeDesign -prePlace \\\n");
  print(FlpFile "-expandedViews -numPaths 1000 -outDir ../RPT/flp
\n\n");
  print(FlpFile "timeDesign -prePlace \\\n");
  print(FlpFile "-hold -expandedViews -numPaths 1000 -outDir ../
RPT/flp \n\n");


  print(FlpFile "summaryReport -noHtml \\\n");
  print(FlpFile "-outfile ../RPT/flp/flp_summaryReport.rpt\n\n");

  print(FlpFile "saveDesign ../$Ldb/flp.enc -compress\n\n");


  print(FlpFile "if { [info exists env(FE_EXIT)] && $env(FE_EXIT)
== 1 } {\n");
  print(FlpFile "  exit \n");
  print(FlpFile "} \n");

  close(FlpFile);
}
else {print "Not building FLP tcl file --> $Output File
exists\n\n";}

### Design Placement plc.tcl ###
$Output       =       $Path.'/'.$Project.'/implementation/physical/
TCL/'.$TopLevelName.'_plc.tcl';
if(!(-e $Output)) {
  print "Building PLACE tcl file --> $Output\n\n";
  unless (open(PlcFile,">$Output"))
    {die "ERROR: could not create PLACE tcl file : $Output\n";}

  print(PlcFile "### Placement Setup Environment ###\n");
  print(PlcFile "\n");
   print(PlcFile "source $Path/$Project/implementation/physical/
TCL/\\\n");
  print(PlcFile "${TopLevelName}_config.tcl\n");
  print(PlcFile "\n");
  print(PlcFile "\n");

  print(PlcFile "### Placement setup ###\n");
  print(PlcFile "\n");

   print(PlcFile "restoreDesign ../$Ldb/flp.enc.dat $TopLevelName
\n");
  print(PlcFile "\n");
```

```
print(PlcFile "source $Path/$Project/physical/TCL/\\\n");
print(PlcFile "${TopLevelName}_setting.tcl\n\n");
print(PlcFile "generateVias\n\n");

print(PlcFile "set_interactive_constraint_modes\\\n");
print(PlcFile " [all_constraint_modes -active]\n");

print(PlcFile "set_max_fanout 50 [current_design]\n");
print(PlcFile "set_max_capacitance 0.300 [current_design]\n");
print(PlcFile "set_max_transition 0.300 [current_design]\n\n");

 print(PlcFile  "update_constraint_mode  -name  setup_func_mode
\\\n");
     print(PlcFile    "          -sdc_files   [list   $Path/$Project/
physical/SDC/\\\n");
 print(PlcFile "${TopLevelName}_func.sdc]\n");
 print(PlcFile "create_analysis_view -name setup_func \\\n");
 print(PlcFile "  -constraint_mode setup_func_mode \\\n");
 print(PlcFile "  -delay_corner slow_max\n\n");

  print(PlcFile  "update_constraint_mode  -name  hold_func_mode
\\\n");
     print(PlcFile    "          -sdc_files   [list   $Path/$Project/
physical/SDC/\\\n");
 print(PlcFile "${TopLevelName}_func.sdc]\n");
 print(PlcFile "create_analysis_view -name hold_func \\\n");
 print(PlcFile "  -constraint_mode hold_func_mode \\\n");
 print(PlcFile "  -delay_corner fast_min\n\n");

 print(PlcFile "set_interactive_constraint_modes [all_constraint_
modes -active]\n");
 print(PlcFile "source $Path/$Project/physical/TCL/\\\n");
 print(PlcFile "${TopLevelName}_clk_gate_disable.tcl\n\n");

 print(PlcFile "set_analysis_view -setup [list setup_func] -hold
[list hold_func]\n\n");


 print(PlcFile "#setScanReorderMode -keepPDPorts true -scanEffort
high\n");
 print(PlcFile "#defIn $Path/$Project/physical/def/\\\n");
 print(PlcFile "${TopLevelName}_scan.def\n\n");

 print(PlcFile "setPlaceMode \\\n");
 print(PlcFile "   -wireLenOptEffort medium \\\n");
```

```perl
print(PlcFile "     -uniformDensity true \\\n");
print(PlcFile "     -maxDensity -1 \\\n");
print(PlcFile "     -placeIoPins false \\\n");
print(PlcFile "     -congEffort auto \\\n");
print(PlcFile "     -reorderScan true \\\n");
print(PlcFile "     -timingDriven true \\\n");
print(PlcFile "     -clusterMode true \\\n");
print(PlcFile "     -clkGateAware true \\\n");
print(PlcFile "     -fp false \\\n");
print(PlcFile "     -ignoreScan false \\\n");
print(PlcFile "     -groupFlopToGate auto \\\n");
print(PlcFile "     -groupFlopToGateHalfPerim 20 \\\n");
print(PlcFile "     -groupFlopToGateMaxFanout 20\n\n");

print(PlcFile "setTrialRouteMode -maxRouteLayer $MaxLayer \n");


print(PlcFile "setOptMode \\\n");
print(PlcFile "     -effort high \\\n");
print(PlcFile "     -preserveAssertions false \\\n");
print(PlcFile "     -leakagePowerEffort none \\\n");
print(PlcFile "     -dynamicPowerEffort none \\\n");
print(PlcFile "     -clkGateAware true \\\n");
print(PlcFile "     -addInst true \\\n");
print(PlcFile "     -allEndPoints true \\\n");
print(PlcFile "     -usefulSkew false \\\n");
print(PlcFile "     -addInstancePrefix PLCOPT_ \\\n");
print(PlcFile "     -fixFanoutLoad true \\\n");
print(PlcFile "     -maxLength 600 \\\n");
print(PlcFile "     -reclaimArea true\n\n");

print(PlcFile "#setOptMode -keepPort ../keep_ports.list\n\n");
print(PlcFile "createClockTreeSpec \\\n");
print(PlcFile "   -bufferList { \\\n");
for ($i=0; $i<@CTS_inv_lst; $i++){
   print(PlcFile "       $CTS_inv_lst[$i] \\\n");}
for ($i=0; $i<@CTS_buf_lst; $i++){
   if ($i != $#CTS_buf_lst) {print(PlcFile "       $CTS_buf_lst[$i]
\\\n");}
   else {print(PlcFile "       $CTS_buf_lst[$i]} \\\n");}
   }
print(PlcFile "   -file ${TopLevelName}_plc.spec \n\n");

print(PlcFile "cleanupSpecifyClockTree\n");
 print(PlcFile "specifyClockTree -file ${TopLevelName}_plc.spec
```

```
\n\n");
  print(PlcFile "### Adding Spare Cells Cluster###\n");
  print(PlcFile "\n");

  print(PlcFile "set nonSpareList [dbGet [dbGet -p \\\n");
   print(PlcFile "[dbGet -p -v top.insts.isSpareGate 1].pstatus
unplaced].name]\n");
    print(PlcFile "foreach i \$nonSpareList {placeInstance \$i
 0 0 -fixed}\n\n");

  print(PlcFile "placeDesign\n\n");


  print(PlcFile "dbDeleteTrialRoute\n\n");


  print(PlcFile "createSpareModule \\\n");
  print(PlcFile "  -moduleName SPARE \\\n");
  print(PlcFile "  -cell {");
  for ($i=0; $i<@SpareCells; $i++){print(PlcFile "$SpareCells[$i]
")}
  print(PlcFile "} \\\n");
  print(PlcFile "  -useCellAsPrefix \n\n");

  print(PlcFile "placeSpareModule \\\n");
  print(PlcFile "  -moduleName SPARE \\\n");
  print(PlcFile "  -prefix SPARE \\\n");
  print(PlcFile "  -stepx 500 \\\n");
  print(PlcFile "  -stepy 500 \\\n");
  print(PlcFile "  -util 0.8 \n\n");

  if ($PwrMgt) {
     print(PlcFile "placeSpareModule \\\n");
     print(PlcFile "  -moduleName SPARE \\\n");
     print(PlcFile "  -prefix SPARE \\\n");
     print(PlcFile "  -stepx 500 \\\n");
     print(PlcFile "  -stepy 500 \\\n");
     print(PlcFile "  -powerDomain PDWN \\\n");
     print(PlcFile "  -util 0.8 \n\n");
  }

  print(PlcFile "### fix spare cells ###\n");
  print(PlcFile "set spareList [dbGet [dbGet -p top.insts.isSpare-
```

```
Gate 1].name]\n");
  print(PlcFile "foreach i \$spareList {dbSet \\\n");
  print(PlcFile "[dbGet -p top.insts.name \$i].pstatus fixed}\n\n");

  print(PlcFile "### unplace non spare cells ###\n");
  print(PlcFile "foreach i \$nonSpareList {dbSet \\\n");
    print(PlcFile  "[dbGet  -p  top.insts.name  \$i].pstatus
unplaced}\n\n");

  print(PlcFile "### Placement ###\n");
  print(PlcFile "\n");

  print(PlcFile "placeDesign -inPlaceOpt\n\n");
  print(PlcFile "timeDesign -prects \\\n");
  print(PlcFile " -prefix PLC0 -expandedViews \\\n");
  print(PlcFile "-numPaths 1000 -outDir ../RPT/plc \n");
  print(PlcFile "\n");
  print(PlcFile "saveDesign ../$Ldb/plc0.enc -compress\n\n");

  print(PlcFile "### Pre-CTS ###\n");
  print(PlcFile "\n");
  print(PlcFile "set_interactive_constraint_modes [all_constraint_
modes -active] \n");
  print(PlcFile "setOptMode -addInstancePrefix PRE_func_ \n");
  print(PlcFile "setCTSMode -clusterMaxFanout 20 \n\n");

  print(PlcFile "optDesign -preCTS \n\n");


  print(PlcFile "clearDrc \n\n");


  print(PlcFile "verifyConnectivity \\\n");
  print(PlcFile "   -type special \\\n");
  print(PlcFile "   -noAntenna \\\n");
  print(PlcFile "   -nets { VSS VDD } \\\n");
  print(PlcFile "   -report ../RPT/plc/plc_conn.rpt \n\n");

  print(PlcFile "verifyGeometry  \\\n");
  print(PlcFile "    -allowPadstd_filerCellsOverlap \\\n");
  print(PlcFile "    -allowRoutingBlkgPinOverlap \\\n");
  print(PlcFile "    -allowRoutingCellBlkgOverlap \\\n");
  print(PlcFile "    -error 1000 \\\n");
  print(PlcFile "    -report ../RPT/plc/plc_geom.rpt\n\n");
```

```
  print(PlcFile "clearDrc \n\n");


  print(PlcFile "setPlaceMode -clkGateAware false\n");
  print(PlcFile "setOptMode -clkGateAware false\n\n");

  print(PlcFile "### To insert buffer tree on scan clock or mbist
clock ###\n");
  print(PlcFile "bufferTreeSynthesis -net dig_top/scanclk -bufList
{ \\\n");
  for ($i=0; $i<@CTS_buf_lst; $i++){
    if ($i != $#CTS_buf_lst) {print(PlcFile "     $CTS_buf_lst[$i]
\\\n");}
    else {print(PlcFile "     $CTS_buf_lst[$i]} \\\n");}
  }
  print(PlcFile "     -maxSkew 100ps -maxFanout 10\n\n");

  print(PlcFile "set_dont_touch [get_cells {dig_top/scanclk__L*}]
true\n");
  print(PlcFile "set_dont_touch [get_nets {dig_top/scanclk__L*}]
true\n\n");

  print(PlcFile "timeDesign -preCTS  -prefix PLC \\\n");
  print(PlcFile " -expandedViews -numPaths 1000 -outDir ../RPT/
plc\n\n");

  print(PlcFile "saveDesign -tcon ../$Ldb/plc.enc -compress\n\n");


  print(PlcFile "summaryReport -noHtml \\\n");
  print(PlcFile "-outfile ../RPT/plc/plc_summaryReport.rpt\n\n");

  print(PlcFile"### Floorplan Specific Wire Load Model ###\n");
  print(PlcFile "#wireload -outfile ../wire_load/\\\n");
    print(PlcFile "${TopLevelName}_wlm  -percent 1.0  -cellLimit
100000\n\n");

  print(PlcFile "if { [info exists env(FE_EXIT)] && \$env(FE_EXIT)
== 1 } {exit}\n");


  close(PlcFile);
}
else {print "Not building PLACE tcl file --> $Output File
exists\n\n";}
```

```
### Design Clock Tree Synthesis cts.tcl ###
$Output      =       $Path.'/'.$Project.'/implementation/physical/
pnrtcl/'.$TopLevelName.'_cts.tcl';
if(!(-e $Output)) {
  print "Building CLOCK SYNTHESIS tcl file --> $Output\n\n";
  unless (open(CtsFile,">$Output"))
    {die "ERROR: could not create CTS tcl file : $Output\n";}

  print(CtsFile "### Clock Tree Synthesis Setup Environment
###\n");
  print(CtsFile "\n");

  print(CtsFile "source $Path/$Project/implementation/physical/
TCL/\\\n");
  print(CtsFile "${TopLevelName}_config.tcl\n");
  print(CtsFile "\n");

  print(CtsFile "### Clock Tree Synthesis Setup ###\n");
  print(CtsFile "\n");

  print(CtsFile "restoreDesign ../$Ldb/plc.enc.dat $TopLevelName
\n");
  print(CtsFile "\n");

  print(CtsFile "source $Path/$Project/implementation/physical/
TCL/\\\n");
  print(CtsFile "${TopLevelName}_setting.tcl\n\n");
  print(CtsFile "generateVias\n\n");

 print(CtsFile "set_interactive_constraint_modes [all_constraint_
modes -active]\n");


  print(CtsFile "update_constraint_mode -name  setup_func_mode
\\\n");
    print(CtsFile "          -sdc_files  [list  $Path/$Project/
physical/SDC/\\\n");
  print(CtsFile "${TopLevelName}_func.sdc]\n");
  print(CtsFile "create_analysis_view -name setup_func \\\n");
  print(CtsFile "   -constraint_mode setup_func_mode \\\n");
  print(CtsFile "   -delay_corner slow_max\n\n");
    print(CtsFile "update_constraint_mode -name  hold_func_mode
\\\n");
    print(CtsFile "          -sdc_files  [list  $Path/$Project/
```

```
physical/SDC/\\\n");
  print(CtsFile "${TopLevelName}_func.sdc]\n");
  print(CtsFile "create_analysis_view -name hold_func \\\n");
  print(CtsFile "    -constraint_mode hold_func_mode \\\n");
  print(CtsFile "    -delay_corner fast_min\n\n");

  print(CtsFile "set_analysis_view -setup\\\n");
  print(CtsFile " [list setup_func] -hold [list hold_func]\n\n");

 print(CtsFile "set_interactive_constraint_modes [all_constraint_
modes -active]\n");


  print(CtsFile "source $Path/$Project/physical/TCL/\\\n");
  print(CtsFile "${TopLevelName}_clk_gate_disable.tcl\n\n");

  print(CtsFile "createClockTreeSpec \\\n");
  print(CtsFile "    -bufferList { \\\n");
  for ($i=0; $i<@CTS_inv_lst; $i++){
    print(CtsFile "     $CTS_inv_lst[$i] \\\n");}
  for ($i=0; $i<@CTS_buf_lst; $i++){
    if ($i != $#CTS_buf_lst) {print(CtsFile "      $CTS_buf_lst[$i]
\\\n");}
    else {print(CtsFile "      $CTS_buf_lst[$i]} \\\n");}}
  print(CtsFile "   -file ${TopLevelName}_clock.spec \n\n");

  print(CtsFile "cleanupSpecifyClockTree\n");
  print(CtsFile "specifyClockTree -file ${TopLevelName}_clock.spec
\n\n");

 print(CtsFile "set_interactive_constraint_modes [all_constraint_
modes -active]\n");


  print(CtsFile "set_max_fanout 50 [current_design]\n");
  print(CtsFile "set_max_capacitance 0.300 [current_design]\n");
  print(CtsFile "set_clock_transition 0.300 [all_clocks]\n\n");

 print(CtsFile "setAnalysisMode -analysisType onChipVariation -cppr
both\n\n");


  print(CtsFile "setNanoRouteMode \\\n");
  print(CtsFile "  -routeWithLithoDriven false \\\n");
  print(CtsFile "  -routeBottomRoutingLayer 1 \\\n");
```

```
  print(CtsFile "    -routeTopRoutingLayer 5 \n\n");

  print(CtsFile "setCTSMode \\\n");
  print(CtsFile "    -clusterMaxFanout 20 \\\n");
  print(CtsFile "    -routeClkNet true \\\n");
  print(CtsFile "    -rcCorrelationAutoMode true \\\n");
  print(CtsFile "    -routeNonDefaultRule NDR_CLK \\\n");
  print(CtsFile "    -useLibMaxCap false \\\n");
  print(CtsFile "    -useLibMaxFanout false \n\n");

  print(CtsFile "set_ccopt_mode \\\n");
  print(CtsFile "    -cts_inverter_cells { \\\n");
  for ($i=0; $i<@CTS_inv_lst; $i++){
     if ($i != $#CTS_inv_lst) {print(CtsFile "    $CTS_inv_lst[$i]
\\\n");}
     else {print(CtsFile "      $CTS_inv_lst[$i]} \\\n");}}
  print(CtsFile "    -cts_buffer_cells { \\\n");
  for ($i=0; $i<@CTS_buf_lst; $i++){
     if ($i != $#CTS_buf_lst) {print(CtsFile "    $CTS_buf_lst[$i]
\\\n");}
     else {print(CtsFile "      $CTS_buf_lst[$i]} \\\n");}}
  print(CtsFile "    -cts_use_inverters true \\\n");
  print(CtsFile "    -cts_target_skew 0.20 \\\n");
  print(CtsFile "    -integration native\n\n");

  print(CtsFile "#set modules [get_cells -filter\\\n");

print(CtsFile "     \"is_hierarchical    ==    true\"    dig_top/
clk_gen/*]\n");
  print(CtsFile "#getReport {query_objects \\\n");
  print(CtsFile "\$modules -limit 10000} > ../keep_ports.list\n");
  print(CtsFile "#setOptMode -keepPort ../keep_ports.list\n\n");

  print(CtsFile "set_interactive_constraint_modes [all_constraint_
modes -active]\n");
  print(CtsFile "set_propagated_clock [all_clocks]\n\n");


print(CtsFile "set restore [get_global timing_defer_mmmc_object_
updates]\n");
     print(CtsFile "set_global  timing_defer_mmmc_object_updates
true\n");
  print(CtsFile "set_analysis_view -update_timing\n");
     print(CtsFile "set_global  timing_defer_mmmc_object_updates
\$restore \n\n");
```

```
print(CtsFile "\");
print(CtsFile "### CTS Stage ###\n");
print(CtsFile "\n");
print(CtsFile "source $Path/$Project/physical/TCL/\\\n");
print(CtsFile "${TopLevelName}_ignore_pins.tcl\n\n");

print(CtsFile "#To keep spares from getting moved\n");
    print(CtsFile  "set_ccopt_property  change_fences_to_guides
false\n\n");

print(CtsFile "set_ccopt_property max_fanout 30\n\n");
print(CtsFile "\n");

print(CtsFile "### for controlling useful skew to help \\\n");
print(CtsFile "hold try these after all ignore pins are found
###\n");
print(CtsFile "### keeps from having too long clock sinks by
\\\n");
print(CtsFile "limiting max insertion delay to no more than 5%
more than avg ### \n");
print(CtsFile "#set_ccopt_property auto_limit_insertion_delay_
factor 1.05 ###\n");
print(CtsFile "### keeps from having too short clock \\\n");
print(CtsFile "sinks by specifying range of skew\n");
print(CtsFile "#set_ccopt_property -target_skew 0.2\n");
print(CtsFile "### forces ccopt to use these constraints ###\n");
print(CtsFile "#set_ccopt_property -constrains ccopt \n\n");

print(CtsFile "create_ccopt_clock_tree_spec -immediate\n\n");


print(CtsFile "ccoptDesign\n\n");


print(CtsFile "timeDesign -expandedViews -numPaths 1000 \\\n");
print(CtsFile " -postCTS -outDir ../RPT/cts -prefix CTS0\n");
print(CtsFile "\n");

print(CtsFile "saveDesign ../$Ldb/cts0.enc -compress\n\n");


print(CtsFile "summaryReport -noHtml -outfile\\\n");
print(CtsFile " ../RPT/cts/cts0_summaryReport.rpt\n\n");
print(CtsFile "\n");
```

```
print(CtsFile "### Post-CTS ###\n");
print(CtsFile "\n");

print(CtsFile "source $Path/$Project/physical/TCL/\\\n");
print(CtsFile "${TopLevelName}_reset_ignore_pins.tcl\n\n");

print(CtsFile "set_interactive_constraint_modes [all_constraint_
modes -active]\n");

print(CtsFile "set_propagated_clock [all_clocks]\n\n");

print(CtsFile "set_interactive_constraint_modes [all_constraint_
modes -active]\n");

print(CtsFile "source $Path/$Project/physical/TCL/\\\n");
print(CtsFile "${TopLevelName}_clk_gate_disable.tcl\n\n");

print(CtsFile "set restore [get_global timing_defer_mmmc_object_
updates]\n");
    print(CtsFile "set_global  timing_defer_mmmc_object_updates
true\n");
  print(CtsFile "set_analysis_view -update_timing\n");
    print(CtsFile "set_global  timing_defer_mmmc_object_updates
\$restore\n\n");

print(CtsFile "setOptMode -fixFanoutLoad true\n");
print(CtsFile "setOptMode -addInstancePrefix CTS1_ \n\n");

print(CtsFile "optDesign -postCTS \n\n");
  print(CtsFile "timeDesign -postCTS -expandedViews -numPaths
1000 -outDir\\\n");
  print(CtsFile " ../RPT/cts -prefix CTS1\n\n");

print(CtsFile "saveDesign ../$Ldb/cts1.enc -compress\n\n");
print(CtsFile "\n");

print(CtsFile "### Switch to Functional SDC ###\n");
print(CtsFile "\n");
print(CtsFile "set_interactive_constraint_modes [all_constraint_
modes -active]\n");
print(CtsFile "cleanupSpecifyClockTree\n\n");
```

```
   print(CtsFile "update_constraint_mode -name setup_func_mode
\\\n");
     print(CtsFile "         -sdc_files [list $Path/$Project/
physical/SDC/\\\n");
  print(CtsFile "${TopLevelName}_func.sdc]\n");
  print(CtsFile "create_analysis_view -name setup_func \\\n");
  print(CtsFile "   -constraint_mode setup_func_mode \\\n");
  print(CtsFile "   -delay_corner slow_max\n\n");

   print(CtsFile "update_constraint_mode -name hold_func_mode
\\\n");
     print(CtsFile "         -sdc_files [list $Path/$Project/
physical/SDC/\\\n");
  print(CtsFile "${TopLevelName}_func.sdc]\n");
  print(CtsFile "create_analysis_view -name hold_func\\\n");
  print(CtsFile "   -constraint_mode hold_func_mode \\\n");
  print(CtsFile "   -delay_corner fast_min\n\n");

  print(CtsFile "set_analysis_view -setup\\\n");
  print(CtsFile " [list setup_func] -hold [list hold_func]\n\n");

 print(CtsFile "set_interactive_constraint_modes [all_constraint_
modes -active]\n");
 print(CtsFile "set_propagated_clock [all_clocks]\n");
 print(CtsFile "source $Path/$Project/physical/TCL/\\\n");
 print(CtsFile "${TopLevelName}_clk_gate_disable.tcl\n\n");


print(CtsFile "set restore [get_global timing_defer_mmmc_object_
updates]\n");
    print(CtsFile "set_global timing_defer_mmmc_object_updates
true\n");
  print(CtsFile "set_analysis_view -update_timing\n");
    print(CtsFile "set_global timing_defer_mmmc_object_updates
\$restore\n\n");

  print(CtsFile "setOptMode -fixFanoutLoad true\n");
  print(CtsFile "setOptMode -addInstancePrefix CTS2_\n\n");
  print(CtsFile "optDesign -postCTS\n\n");

   print(CtsFile "timeDesign -postCTS -expandedViews -numPaths
1000\\\n");
  print(CtsFile " -outDir ../RPT/cts -prefix CTS2\n");
  print(CtsFile "timeDesign -postCTS -hold -expandedViews -numP-
aths 1000 \\\n");
```

```
  print(CtsFile "-outDir ../RPT/cts -prefix CTS2\n\n");

  print(CtsFile "saveDesign ../$Ldb/cts2.enc -compress\n\n");
  print(CtsFile "\n");

  print(CtsFile "### hold fix ###\n");
  print(CtsFile "\n");
  print(CtsFile "set_interactive_constraint_modes [all_constraint_
modes -active]\n");


print(CtsFile "set restore [get_global timing_defer_mmmc_object_
updates]\n");
     print(CtsFile  "set_global  timing_defer_mmmc_object_updates
true\n");
  print(CtsFile "set_analysis_view -update_timing\n");
     print(CtsFile  "set_global  timing_defer_mmmc_object_updates
\$restore \n\n");

  print(CtsFile "setAnalysisMode -honorClockDomains true\n\n");


  print(CtsFile "setOptMode -addInstancePrefix hold_FIX_ \n\n");


  print(CtsFile "set_interactive_constraint_modes [all_constraint_
modes -active]\n");


  print(CtsFile "source $Path/$Project/physical/TCL/\\\n");
  print(CtsFile "${TopLevelName}_clk_gate_disable.tcl\n\n");

  print(CtsFile "setOptMode \\\n");
  print(CtsFile "  -fixHoldAllowSetupTnsDegrade false \\\n");
    print(CtsFile "    -ignorePathGroupsForHold {reg2out in2out}
\n\n");

  for ($i=0; $i<@Hold_buf_lst; $i++){
    print(CtsFile "setDontUse $Hold_buf_lst[$i] false\n");}
  print(CtsFile "\n");
  print(CtsFile "setOptMode -holdFixingCells { \\\n");
  for ($i=0; $i<@Hold_buf_lst; $i++){
    if ($i != $#Hold_buf_lst) {print(CtsFile " $Hold_buf_lst[$i]
\\\n");}
    else {print(CtsFile "     $Hold_buf_lst[$i] }\n\n");} }
```

```
  print(CtsFile "optDesign -postCTS -hold\n\n");


 print(CtsFile "timeDesign -postCTS -expandedViews -numPaths 1000
\\\n");
  print(CtsFile "-outDir ../RPT/cts -prefix CTS\n");
  print(CtsFile "timeDesign -hold -postCTS -expandedViews -numP-
aths 1000 \\\n);
  print(CtsFile "-outDir ../RPT/cts -prefix CTS\n\n");

  print(CtsFile "saveDesign ../$Ldb/cts.enc\n\n");


 print(CtsFile "summaryReport -noHtml \\\n);
  print(CtsFile "-outfile ../RPT/cts/cts_summaryReport.rpt\n\n");

 print(CtsFile "if { [info exists env(FE_EXIT)]·&& \$env(FE_EXIT)
== 1 } {exit}\n");


  close(CtsFile);
}
else {print "Not building CLOCK SYNTHESIS tcl file --> $Output File
exists\n\n";}

### Design Final Route frt.tcl ###
$Output      =      $Path.'/'.$Project.'/implementation/physical/
TCL/'.$TopLevelName.'_frt.tcl';
if(!(-e $Output)) {
  print "Building ROUTING tcl file --> $Output\n\n";
  unless (open(FrtFile,">$Output"))
   {die "ERROR: could not create ROUTING tcl file : $Output\n";}

  print(FrtFile "### Final Route Setup Environment ###\n");
  print(FrtFile "\n");
  print(FrtFile "source $Path/$Project/physical/TCL/\\\n");
  print(FrtFile "${TopLevelName}_config.tcl\n");
  print(FrtFile "\n");

  print(FrtFile "### Final Route setup ###\n");
  print(FrtFile "\n");

 print(FrtFile "restoreDesign ../$Ldb/cts.enc.dat $TopLevelName\n");
  print(FrtFile "\n");
```

```
print(FrtFile "source $Path/$Project/physical/TCL/\\\n");
print(FrtFile "${TopLevelName}_setting.tcl\n\n");

print(FrtFile "generateVias\n\n");


print(FrtFile "set_interactive_constraint_modes [all_constraint_
modes -active]\n\n");


print(FrtFile "set_analysis_view -setup\\\n");
print(FrtFile " [list setup_func] -hold [list hold_func]\n\n");

print(FrtFile "source $Path/$Project/physical/TCL/\\\n");
print(FrtFile "${TopLevelName}_clk_gate_disable.tcl\n\n");

print(FrtFile "setNanoRouteMode -routeTopRoutingLayer $MaxLayer
\n");
    print(FrtFile "setNanoRouteMode -routeWithLithoDriven false
\n");
    print(FrtFile "setNanoRouteMode -routeWithTimingDriven false
\n");
print(FrtFile "setNanoRouteMode -routeWithSiDriven true \n");
print(FrtFile "setNanoRouteMode -droutePostRouteSpreadWire false
\n\n");

print(FrtFile "setSIMode -deltaDelayThreshold 0.01 \\\n");
print(FrtFile "-analyzeNoiseThreshold 80 -fixGlitch false\n\n");

print(FrtFile "setOptMode -addInstancePrefix INT_FRT_\n\n");


print(FrtFile "set active_corners [all_delay_corners]\n");
print(FrtFile "setAnalysisMode -analysisType onChipVariation -cppr
setup \n\n");

print(FrtFile "### Final Route Stage ###\n");
print(FrtFile "\n");
print(FrtFile "routeDesign \n");
print(FrtFile "timeDesign -postRoute -prefix INT_FRT -expanded-
Views -numPaths\\\n");
print(FrtFile " 1000 -outDir ../RPT/frt \n\n");

print(FrtFile "saveDesign ../$Ldb/INT_FRT.enc -compress\n");
print(FrtFile "\n");
```

```
print(FrtFile "### DFM ###\n");
print(FrtFile "#setNanoRouteMode -droutePostRouteSpreadWire true
\\\n)";
print(FrtFile "-routeWithTimingDriven false\n");
print(FrtFile "#routeDesign -wireOpt\n");
   print(FrtFile  "#setNanoRouteMode  -droutePostRouteSwapVia
multiCut\n");
 print(FrtFile "#setNanoRouteMode -drouteMinSlackForWireOptimi-
zation <slack>\n");
print(FrtFile "#routeDesign -viaOpt\n");
  print(FrtFile "#setNanoRouteMode  -droutePostRouteSpreadWire
false\\\n):
print(FrtFile " -routeWithTimingDriven true\n\n");

print(FrtFile "### Route Optimization ###\n");
print(FrtFile "\n");

   print(FrtFile  "setNanoRouteMode  -drouteUseMultiCutViaEffort
medium \n\n");


print(FrtFile "setAnalysisMode -analysisType\\\n");
print(FrtFile " onChipVariation -cppr both \n\n");

  print(FrtFile "setDelayCalMode -SIAware true -engine default
\n\n");


print(FrtFile "setExtractRCMode -engine postRoute\\\n");
print(FrtFile " -coupled true -effortLevel medium\n\n");

 print(FrtFile "set_interactive_constraint_modes [all_constraint_
modes -active]\n");
print(FrtFile "set_propagated_clock [all_clocks] \n\n");
print(FrtFile "setOptMode -addInstancePrefix FRT_ \n\n");
print(FrtFile "optDesign -postRoute -prefix OPT_FRT \n\n");
print(FrtFile "setOptMode -addInstancePrefix FRT_HOLD_ \n\n");

for ($i=0; $i<@Hold_buf_lst; $i++){
  print(FrtFile "setDontUse $Hold_buf_lst[$i] false\n");}
print(FrtFile "\n");

print(FrtFile "setOptMode -holdFixingCells { \\\n");
for ($i=0; $i<@Hold_buf_lst; $i++){
```

```
     if ($i != $#Hold_buf_lst) {print(FrtFile " $Hold_buf_lst[$i]
\\\n");}
     else {print(FrtFile "  $Hold_buf_lst[$i] }\n\n");}
  }


  print(FrtFile "set_interactive_constraint_modes [all_constraint_
modes -active]\n");


  print(FrtFile "source $Path/$Project/physical/TCL/\\\n");
  print(FrtFile "${TopLevelName}_clk_gate_disable.tcl\n\n");

  print(FrtFile "optDesign -postRoute -hold -outDir\\\n");
  print(FrtFile " ./RPT/frt -prefix OPT_FRT_HOLD \n");
  print(FrtFile "timeDesign -postRoute -hold -prefix \\\n");
  print(FrtFile "OPT_FRT_HOLD -expandedViews -numPaths 1000 -out-
Dir ../RPT/frt \n\n");
  print(FrtFile "timeDesign -postRoute -prefix OPT_FRT \\\n");
  print(FrtFile "-expandedViews -numPaths 1000 -outDir ../RPT/frt
\n\n");

  print(FrtFile "saveDesign ../$Ldb/OPT_FRT.enc -compress\n\n");


  print(FrtFile "optDesign -postRoute -outDir\\\n");
  print(FrtFile " ./RPT/frt -prefix FRT\n\n");

  print(FrtFile "### Leakage Optimize ###\n");
  print(FrtFile "\n");

  print(FrtFile "#report_power -leakage \n");
  print(FrtFile "#optLeakagePower \n");
  print(FrtFile "#report_power -leakage \n\n");

#  print(FrtFile "### SI Optimize ###\n");
#  print(FrtFile "\n");
#
#    print(FrtFile  "#set_interactive_constraint_modes  [all_con-
straint_modes -active]\n");

#         print(FrtFile     "#setAnalysisMode    -analysisType
onChipVariation -cppr both\n");
#    print(FrtFile  "#setDelayCalMode  -SIAware  false  -engine
signalstorm\n\n");
```

```
#   print(FrtFile "#setSIMode -fixDRC true -fixDelay true\\\n");
    print(FrtFile " -fixHoldIncludeXtalkSetup true -fixGlitch false
\n\n");


#       print(FrtFile  "#setOptMode   -fixHoldAllowSetupTnsDegrade
false\\\n");
  print(FrtFile " -ignorePathGroupsForHold {reg2out in2out} \n");
  print(FrtFile "-outDir ../RPT/frt -prefix FRT_SI \n\n");
  printFrtFile "\n");

  print(FrtFile "### Finishing ###\n");
  print(FrtFile "\n");

    print(FrtFile  "setNanoRouteMode  -droutePostRouteLithoRepair
false \n\n");



  print(FrtFile "setNanoRouteMode -drouteSearchAndRepair true\n");
  print(FrtFile "globalDetailRoute \n\n");
      print(FrtFile   "timeDesign   -postRoute   -hold   -prefix
FRT_HOLD\\\n");
   print(FrtFile " -expandedViews -numPaths 1000 -outDir ../RPT/
frt\n");
  print(FrtFile "timeDesign -postRoute -prefix FRT \\\n");
   print(FrtFile "-expandedViews -numPaths 1000 -outDir ../RPT/
frt\n\n");

  print(FrtFile "deleteEmptyModule\n\n");



  print(FrtFile "saveDesign -tcon ../$Ldb/frt.enc -compress \n\n");



  print(FrtFile "summaryReport -noHtml -outfile \\\n");
    print(FrtFile   "../RPT/frt/${TopLevelName}_summaryReport.rpt
\n\n");

  print(FrtFile "setFillerMode -corePrefix std_fil -core \"");
  for ($i=0; $i<@Stdstd_fil; $i++){
     if ($i != $#Std_fil) {
        print(FrtFile "$Std_fil[$i] ");}
     else{
        print(FrtFile "$Std_fil[$i]\"\n");}}
  print(FrtFile "addFiller \n\n");
```

```
   print(FrtFile "verifyConnectivity -noAntenna\n");
   print(FrtFile "verifyGeometry\n");
   print(FrtFile "verifyProcessAntenna\n\n");

   print(FrtFile "saveDesign -tcon ../$Ldb/\\\n");
   print(FrtFile "${TopLevelName}.enc -compress \n\n");

   print(FrtFile "if { [info exists env(FE_EXIT)] && \$env(FE_EXIT)
== 1 }
                    {exit}\n");
   close(FrtFile);
}
else {print "Not building ROUTING tcl file --> $Output File
exists\n\n";}


### Design Engineering Change Order ECO.tcl ###
$Output       =      $Path.'/'.$Project.'/implementation/physical/
TCL/'.$TopLevelName.'_eco.tcl';
if(!(-e $Output)) {
   print "Building ECO tcl file --> $Output\n\n";
   unless (open(EcoFile,">$Output"))
      {die "ERROR: could not create ECO tcl file : $Output\n";}

   print(EcoFile "### Engineering Change Order Setup Environment
###\n");
   print(EcoFile "\n");
   print(EcoFile "source $Path/$Project/implementation/physical/
TCL/\\\n");
   print(EcoFile "${TopLevelName}_config.tcl\n");
   print(EcoFile "\n");

   print(EcoFile "restoreDesign ../$Ldb/frt.enc.dat $TopLevelName
\n");
   print(EcoFile "\n");

   print(EcoFile "source $Path/$Project/physical/TCL/\\\n");
   print(EcoFile "${TopLevelName}_setting.tcl\n\n");
   print(EcoFile "generateVias\n\n");

   print(EcoFile "saveDesign ../$Ldb/frt_0.enc -compress \n\n");


   print(EcoFile "### Netlist Based ECO ###\n");
   print(EcoFile "\n");
   print(EcoFile "ecoDesign -noEcoPlace -noEcoRoute ../$Ldb/frt.
```

```
enc.dat\\\n");
  print(EcoFile " $TopLevelName ../ecos/${TopLevelName}_func_eco.
vg \n");


  if ($PadFile eq "") {}
    else { print(EcoFile "\naddInst -cell $AddInst[0] -inst
$AddInst[0]\n\n");}

  print(EcoFile "ecoPlace \n");
  print(EcoFile "\n");

  print(EcoFile "### TCL Based ECO ###\n");
  print(EcoFile "\n");

  print(EcoFile "setEcoMode -honorDontUse false\\\n");
 print(EcoFile " -honorDontTouch false -honorFixedStatus false\n");
  print(EcoFile "setEcoMode -refinePlace false\\\n");
  print(EcoFile " -updateTiming false -batchMode true \n\n");

  print(EcoFile "setOptMode -addInstancePrefix ECO1_\n\n");


  print(EcoFile "source ../ecos/${TopLevelName}_eco.tcl \n");
  print(EcoFile "\n");

  print(EcoFile "refinePlace -preserveRouting true\n");
  print(EcoFile "checkPlace\n\n");
  print(EcoFile "### ECO Routing ###\n");
  print(EcoFile "\n");
    print(EcoFile "setNanoRouteMode -droutePostRouteLithoRepair
false \n");
    print(EcoFile "setNanoRouteMode -routeWithLithoDriven false
\n\n");

  print(EcoFile "### For limiting layers in ECO ###\n");
  print(EcoFile "#setNanoRouteMode -routeEcoOnlyInLayers 1:4\n");
  print(EcoFile "#ecoRoute -modifyOnlyLayers 1:4\n");
  print(EcoFile "#routeDesign\n\n");

  print(EcoFile "### For regular all layer ECO ###\n");
  print(EcoFile "ecoRoute \n");
  print(EcoFile "routeDesign \n\n");

  print(EcoFile "### Antenna Fixing ###\n");
  print(EcoFile "\n");
```

```
print(EcoFile "#To fix remaining antenna violations\\\n");
print(EcoFile " by inserting diodes automatically\n");
print(EcoFile "#setNanoRouteMode -drouteFixAntenna true\\\n");
print(EcoFile " -routeInsertAntennaDiode true -routeAntennaCell-
Name {ADIODE}\n");
print(EcoFile "#routeDesign\n");
print(EcoFile "#setNanoRouteMode -drouteFixAntenna true\\\n");
print(EcoFile " -routeInsertAntennaDiode false\n");
print(EcoFile "#routeDesign\n\n");

print(EcoFile "### To add antenna diode by manual ECO ###\n");
print(EcoFile "#attachDiode -diodeCell ADIODE -pin dig_top/
mclk_scanmux A\n\n");

print(EcoFile "### Excessive Routing Violations ###\n");
print(EcoFile "\n");

print(EcoFile "### Strategy: Remove and re-route nets with shorts
###\n");
print(EcoFile "#reroute_shorts\n");
print(EcoFile "#ecoRoute\n");
print(EcoFile "#routeDesign\n\n");

print(EcoFile "### Strategy: Relax clock NDR rules on nets in
congested area ###\n");
print(EcoFile "### list all net names on which to relax con-
straints here ###\n");
print(EcoFile "#set ndr_nets { List of NRD clock nets here}\n");
print(EcoFile "#remove_ndr_nets \$ndr_nets\n");
print(EcoFile "#ecoRoute\n");
print(EcoFile "#routeDesign\n\n");

print(EcoFile "### Check for missing vias ###\n");
print(EcoFile "verifyPowerVia -layerRange {MET4 MET1}\n\n");

print(EcoFile "saveDesign ../$Ldb/frt.enc -compress\n");
print(EcoFile "\n");

print(EcoFile "setstd_filerMode -corePrefix std_fil -core \"");
for ($i=0; $i<@Stdstd_fil; $i++){
    if ($i != $#Stdstd_fil) {
      print(EcoFile "$Stdstd_fil[$i] ");}
    else{
      print(EcoFile "$Stdstd_fil[$i]\"\n");}}
print(EcoFile "addstd_filer \n\n");
```

```
  print(EcoFile "saveDesign ../$Ldb/${TopLevelName}.enc -compress
\n\n");


  print(EcoFile "if { [info exists env(FE_EXIT)] && \$env(FE_EXIT)
== 1 }
                     {exit}\n");

  close(EcoFile);
}
else  {print  "Not  building  ECO  tcl  file  -->  $Output  File
exists\n\n";}

### Design GDS Export gds.tcl ###
$Output      =        $Path.'/'.$Project.'/implementation/physical/
TCL/'.$TopLevelName.'_gds.tcl';
if(!(-e $Output)) {
  print "Building EXPORT GDS tcl file --> $Output\n\n";
  unless (open(GdsFile,">$Output"))
    {die "ERROR: could not create EXPORT GDS tcl file : $Output\n";}

  print(GdsFile "### Export GDS Setup Environment ###\n");
  print(GdsFile "\n");
   print(GdsFile "source $Path/$Project/implementation/physical/
TCL/\\\n");
  print(GdsFile "${TopLevelName}_config.tcl\n");
  print(GdsFile "\n");

  print(GdsFile "### Export GDS From EDI Database ###\n");
  print(GdsFile "\n");

  print(GdsFile "restoreDesign ../$Ldb/\\\n");
  print(GdsFile "$TopLevelName.enc.dat $TopLevelName \n\n");

  print(GdsFile "source $Path/$Project/physical/TCL/\\\n");
  print(GdsFile "${TopLevelName}_setting.tcl \n");
  print(GdsFile "\n");

  if ($PadFile eq ""){
        print(GdsFile  "streamOut  $Path/$Project/physical/gds/
macros/\\\n");
             print(GdsFile    "${TopLevelName}_edi.gds   -mode
ALL -dieAreaAsBoundary\\\n");
    print(GdsFile " -mapFile ../streamOut.map \n");
    print(GdsFile "\n");}
```

```
  else{
    print(GdsFile "streamOut $Path/$Project/physical/gds/\\\n");
                 print(GdsFile    "${TopLevelName}_edi.gds    -mode
ALL -dieAreaAsBoundary\\\n");
    print(GdsFile " -mapFile ../streamOut.map \n");
    print(GdsFile "\n");}

  print(GdsFile "if { [info exists env(FE_EXIT)] && \$env(FE_EXIT)
== 1 }
                      {exit}\n");
  close(GdsFile);
}
else {print "Not building EXPORT GDS tcl file --> $Output  File
exists\n\n";}


### Export Design Netlist net.tcl ###
$Output = $Path.'/'.$Project.'/implementation/physical/TCL/\\\n");
  print(NetFile "'.$TopLevelName.'_net.tcl';
if(!(-e $Output)) {
  print "Building OUTPUT VERILOG tcl file --> $Output\n\n";
  unless (open(NetFile,">$Output"))
      {die "ERROR: could not create OUTPUT VERILOG tcl file :
$Output\n";}

  print(NetFile "### Export Netlist Setup Environment ###\n");
  print(NetFile "\n");
   print(NetFile "source $Path/$Project/implementation/physical/
TCL/\\\n");
  print(NetFile "${TopLevelName}_config.tcl\n");
  print(NetFile "\n");
  print(NetFile "### Export Verilog Netlist ###\n");
  print(NetFile "\n");

  print(NetFile "restoreDesign ../$Ldb/\\\n");
  print(NetFile "${TopLevelName}.enc.dat $TopLevelName \n");
  print(NetFile "\n");

  print(NetFile "source $Path/$Project/physical/TCL/\\\n");
  print(NetFile "${TopLevelName}_setting.tcl\n");
  print(NetFile "\n");

  print(NetFile "### write no Power/Ground netlist ###\n");
  print(NetFile "saveNetlist $Path/$Project/physical/net/\\\n");
   print(NetFile "${TopLevelName}_func.vg -topCell $TopLevelName
\\\n");
```

```
print(NetFile "                -excludeLeafCell \\\n");
print(NetFile "          -excludeTopCellPGPort {VDD VSS} \\\n");
print(NetFile "              -excludeCellInst {$AddInst[0]} \n");
print(NetFile "\n");

print(NetFile "### write with Power/Ground netlist\n");
print(NetFile "saveNetlist $Path/$Project/physical/net/\\\n");
print(NetFile "${TopLevelName}_func_pg.vg \\\n");
print(NetFile "                -excludeLeafCell \\\n");
print(NetFile "                -includePowerGround \\\n");
print(NetFile "              -excludeCellInst {$AddInst[0]} \n");
print(NetFile "\n");

print(NetFile"### write DEF ###\n");
print(NetFile "\n");

    print(NetFile  "defOut  -routing  $Path/$Project/physical/
def/\\\n");
print(NetFile "${TopLevelName}.def \n");
print(NetFile "\n");

print(NetFile "if { [info exists env(FE_EXIT)] && \$env(FE_EXIT)
== 1 }
                {exit}\n");
close(NetFile);
}
else {print "Not building OUTPUT VERILOG tcl file --> $Output File
exists\n\n";}

### Export Design SPEF spef.tcl ###
$Output       =        $Path.'/'.$Project.'/implementation/physical/
TCL/'.$TopLevelName.'_spef.tcl';
if(!(-e $Output)) {
  print "Building OUTPUT SPEF tcl file --> $Output\n\n";
  unless (open(SpefFile,">$Output"))
    {die "ERROR: could not create OUTPUT SPEF tcl file : $Output\n";}

  print(SpefFile "### Export SPEF Setup Environment ###\n");
  print(SpefFile "\n");
  print(SpefFile "source $Path/$Project/implementation/physical/
TCL/\\\n");
  print(SpefFile "${TopLevelName}_config.tcl\n");
  print(SpefFile "\n");
  print(SpefFile "### Export SPEF Files ###\n");
  print(SpefFile "\n");
```

```
   print(SpefFile "restoreDesign ../$Ldb/${TopLevelName}.enc.dat
$TopLevelName\n");
  print(SpefFile "\n");

  print(SpefFile "source $Path/$Project/physical/TCL/\\\n");
  print(SpefFile "${TopLevelName}_setting.tcl\n");
  print(SpefFile "\n");

 print(SpefFile "setExtractRCMode -coupled true -effortLevel\\\n");
  print(SpefFile " high -engine postRoute -coupling_c_th 3\\\n");
  print(SpefFile " -relative_c_th 0.03 -total_c_th 3 \n");
          print(SpefFile     "setAnalysisMode      -analysisType
onChipVariation -cppr both \n\n");

  print(SpefFile "extractRC\n\n");


  print(SpefFile "rcOut -rc_corner rc_max -spef $Path/$Project/
physical/spef/\\\n;
  print(SpefFile "${TopLevelName}_max.spef.gz\n");
  print(SpefFile "rcOut -rc_corner rc_min -spef $Path/$Project/
physical/spef/\\\n");
  print(SpefFile "${TopLevelName}_min.spef.gz\n");
  print(SpefFile "\n");

  print(SpefFile "if { [info exists env(FE_EXIT)] && \$env(FE_
EXIT) == 1 }
                      {exit}\n");
  close(SpefFile);
}
else {print "Not building OUTPUT SPEF tcl file --> $Output File
exists\n\n";}

### Design Multi-Mode Multi-Corner mmmc.tcl ###
$Output      =      $Path.'/'.$Project.'/implementation/physical/
TCL/'.$TopLevelName.'_mmmc.tcl';
if(!(-e $Output)) {
 print "Building MULTI-MODE MULTI-CORNER tcl file --> $Output\n\n";
 unless (open(MmcFile,">$Output"))
    {die "ERROR: could not create ROUTING tcl file : $Output\n";}

  print(MmcFile "### Multi-Mode Multi-Corner Setup Environment
###\n");
 print(MmcFile "\n");
  print(MmcFile "source $Path/$Project/implementation/physical/
```

```
TCL/\\\n");
  print(MmcFile "${TopLevelName}_config.tcl\n");
  print(MmcFile "\n");

  print(MmcFile "### Multi-Mode Multi-Corner Setup ###\n");
  print(MmcFile "\n");

print(MmcFile"restoreDesign../$Ldb/frt.enc.dat $TopLevelName\n");
  print(MmcFile "\n");

  print(MmcFile "source $Path/$Project/physical/TCL/\\\n");
  print(MmcFile "${TopLevelName}_setting.tcl\n\n");

  print(MmcFile "generateVias\n\n");


      print(MmcFile    "saveDesign   ../$Ldb/frt_no_mmc_opt.enc
-compress\n\n");


  print(MmcFile "\n");
    print(MmcFile  "set_analysis_view -setup [list  setup_func
setup_mbist]\\\n");
    print(MmcFile  "-hold  [list  hold_func  hold_scanc  hold_scans
hold_mbist]\n");
  print(MmcFile "\n");
 print(MmcFile "set_interactive_constraint_modes [all_constraint_
modes -active]\n");
  print(MmcFile "\n");
    print(MmcFile  "update_constraint_mode  -name  setup_func_mode
\\\n");
      print(MmcFile   "        -sdc_files   [list   $Path/$Project/
physical/SDC/\\\n");
  print(MmcFile "${TopLevelName}_func.sdc]\n");
  print(MmcFile "create_analysis_view -name setup_func \\\n");
  print(MmcFile "  -constraint_mode setup_func_mode \\\n");
  print(MmcFile "  -delay_corner slow_max\n");
  print(MmcFile "\n");
    print(MmcFile  "update_constraint_mode  -name  hold_func_mode
\\\n");
      print(MmcFile   "        -sdc_files   [list   $Path/$Project/
physical/SDC/\\\n");
  print(MmcFile "${TopLevelName}_func.sdc]\n");
  print(MmcFile "create_analysis_view -name hold_func\\\n");
  print(MmcFile "  -constraint_mode hold_func_mode \\\n");
```

```
  print(MmcFile "   -delay_corner fast_min\n");
  print(MmcFile "\n");
    print(MmcFile "update_constraint_mode -name hold_scans_mode
\\\n");
       print(MmcFile   "        -sdc_files  [list  $Path/$Project/
physical/SDC/\\\n");
  print(MmcFile "${TopLevelName}_scans.sdc]\n");
  print(MmcFile "create_analysis_view -name hold_scans \\\n");
  print(MmcFile "   -constraint_mode hold_scans_mode \\\n");
  print(MmcFile "   -delay_corner fast_min\n");
  print(MmcFile "\n");
    print(MmcFile "update_constraint_mode -name hold_scanc_mode
\\\n");
       print(MmcFile   "        -sdc_files  [list  $Path/$Project/
physical/SDC/\\\n");
  print(MmcFile "${TopLevelName}_scanc.sdc]\n");
  print(MmcFile "create_analysis_view -name hold_scanc \\\n");
  print(MmcFile "   -constraint_mode hold_scanc_mode \\\n");
  print(MmcFile "   -delay_corner fast_min\n");
  print(MmcFile "\n");
    print(MmcFile "update_constraint_mode -name setup_mbist_mode
\\\n");
     print(MmcFile   "        -sdc_files  [list  $Path/$Project/
physical/SDC/\\\n");
  print(MmcFile "${TopLevelName}_mbist.sdc]\n");
  print(MmcFile "create_analysis_view -name setup_mbist \\\n");
  print(MmcFile "   -constraint_mode setup_mbist_mode \\\n");
  print(MmcFile "   -delay_corner slow_max\n");
  print(MmcFile "\n");
    print(MmcFile "update_constraint_mode -name hold_mbist_mode
\\\n");
       print(MmcFile   "        -sdc_files  [list  $Path/$Project/
physical/SDC/\\\n");
  print(MmcFile "${TopLevelName}_mbist.sdc]\n");
  print(MmcFile "create_analysis_view -name hold_mbist \\\n");
  print(MmcFile "   -constraint_mode hold_mbist_mode \\\n");
  print(MmcFile "   -delay_corner fast_min\n");
  print(MmcFile "\n");

  print(MmcFile "set_analysis_view \\\n");
  print(MmcFile "   -setup [list setup_func setup_mbist] \\\n");
    print(MmcFile"-hold [list  hold_func hold_scanc hold_scans
hold_mbist]\n");
  print(MmcFile "\n");
 print(MmcFile "set_interactive_constraint_modes [all_constraint_
```

```
modes -active]\n");
  print(MmcFile "source $Path/$Project/physical/TCL/\\\n");
  print(MmcFile "${TopLevelName}_clk_gate_disable.tcl\n\n");


 print(MmcFile "set_interactive_constraint_modes [all_constraint_
modes -active]\n");
  print(MmcFile "set report_timing_format\\\n");
    print(MmcFile "  {instance cell pin arc fanout load delay
arrival}\n");
  print(MmcFile "set_propagated_clock [all_clocks]\n");
  print(MmcFile "\n");
  print(MmcFile "timeDesign -postRoute -numPaths 1000 \\\n");
  print(MmcFile "-outDir ../RPT/pre_mmc -expandedViews\n");
 print(MmcFile "timeDesign -hold -postRoute -numPaths 1000 \\\n");
  print(MmcFile "-outDir ../RPT/pre_mmc -expandedViews\n");
  print(MmcFile "\n");
  print(MmcFile "setOptMode -addInstancePrefix MMC_\n\n");
  print(MmcFile "optDesign -postRoute\n");
  print(MmcFile "timeDesign -postRoute -numPaths 1000 \\\n");
  print(MmcFile "-outDir ../RPT/mmc_no_hold -expandedViews\n");
 print(MmcFile "timeDesign -hold -postRoute -numPaths 1000 \\\n");
  print(MmcFile "-outDir ../RPT/mmc_no_hold -expandedViews\n");
  print(MmcFile "\n");
    print(MmcFile  "saveDesign  -tcon  ../$Ldb/frt_mmc_no_hold.
enc -compress\n\n");


  for ($i=0; $i<@Hold_buf_lst; $i++){
    print(MmcFile "setDontUse $Hold_buf_lst[$i] false\n");}
  print(MmcFile "\n");


  print(MmcFile "setOptMode -holdFixingCells { \\\n");
  for ($i=0; $i<@Hold_buf_lst; $i++){
     if ($i != $#Hold_buf_lst) {print(MmcFile "       $Hold_buf_
lst[$i] \\\n");}
    else {print(MmcFile "      $Hold_buf_lst[$i] }\n\n");}
   }


       print(MmcFile  "setOptMode  -fixHoldAllowSetupTnsDegrade
false\\\n");
  print(MmcFile " -ignorePathGroupsForHold {reg2out in2out}\n");
  print(MmcFile "\n");
  print(MmcFile "optDesign -hold -postroute\n");
  print(MmcFile "optDesign -postRoute\n");
  print(MmcFile "timeDesign -postRoute -hold \\\n");
 print(MmcFile "-outDir ../RPT/mmc -numPaths 1000 -expandedViews\n");
```

```perl
  print(MmcFile "timeDesign -postRoute \\\n");
 print(MmcFile "-outDir ../RPT/mmc -numPaths 1000 -expandedViews\n");
  print(MmcFile "\n");
  print(MmcFile "saveDesign -tcon ../$Ldb/frt.enc -compress\n");
  print(MmcFile "\n");

  print(MmcFile "summaryReport -noHtml \\\n");
  print(MmcFile "-outfile ../RPT/mmc/\\\n");
  print(MmcFile "${TopLevelName}_summaryReport.rpt \n\n");

  print(MmcFile "setstd_filerMode -corePrefix std_fil -core \"");
  for ($i=0; $i<@Stdstd_fil; $i++){
     if ($i != $#Stdstd_fil) {
       print(MmcFile "$Stdstd_fil[$i] ");}
     else{
       print(MmcFile "$Stdstd_fil[$i]\"\n");}}
  print(MmcFile "addstd_filer \n\n");

     print(MmcFile  "saveDesign  -tcon  ../$Ldb/$TopLevelName.
enc -compress\n");
  print(MmcFile "\n");
}
else {print "Not building MULTI_MODE_MULTI_CORNER tcl file -->
Design File exists\n\n";}

### Design Makefile runAll.tcl ###
$Output = $Path.'/'.$Project.'/implementation/physical/PNR. $'/
runAll.tcl';
if(!(-e $Output)) {
  unless(open(runAllFile,">$Output"))
    { die "ERROR: could not write to file $Output\n"; }
  print(runAllFile "\#!/bin/tcsh -f \n\n");
  print(runAllFile "setcnv FE_EXIT 1 \n\n");
   print(runAllFile "encounter -64 -log ../logs/flp.log -replay
../../tcl/\\\n");
  print(runAllFile "${TopLevelName}_flp.tcl\n");
   print(runAllFile "encounter -64 -log ../logs/plc.log -replay
../../tcl/\\\n");
  print(runAllFile "${TopLevelName}_plc.tcl\n");
   print(runAllFile "encounter -64 -log ../logs/cts.log -replay
../../tcl/\\\n");
  print(runAllFile "${TopLevelName}_func.tcl\n");
   print(runAllFile "encounter -64 -log ../logs/frt.log -replay
../../tcl/\\\n");
  print(runAllFile "${TopLevelName}_frt.tcl\n");
```

```perl
    print(runAllFile "encounter -64 -log ../logs/mmc.log -replay
../../tcl/\\\n");
  print(runAllFile "${TopLevelName}_mmc.tcl\n");
   print(runAllFile "encounter -64 -log ../logs/net.log -replay
../../tcl/\\\n");
  print(runAllFile "${TopLevelName}_net.tcl\n");
   print(runAllFile "encounter -64 -log ../logs/spef.log -replay
../../tcl/\\\n");
  print(runAllFile "${TopLevelName}_spef.tcl\n");
   print(runAllFile "encounter -64 -log ../logs/gds.log -replay
../../tcl/\\\n");
  print(runAllFile "${TopLevelName}_gds.tcl\n");
  close(runAllFile);
}
sub parse_command_line {
    for ($i=0; $i<=$#ARGV; $i++)
      { $_ = $ARGV[$i];
        if (/^-i/) { $DataFile = $ARGV[++$i] }
        if (/^-h\b/) { &print_usage }
        }
    unless ( $DataFile )
      {
        print "\n";
        print "ERROR: No options specified.\n";
        print "\nusage: make_pd_tcl -i design_pd.env.dat\n";
        print "\n";
        exit(0); }
}
sub print_usage {
    print "\nusage: make_pd_tcl -i design_pd.envdat\n";
    print "\n";
    print " -i  #design_pd.env is project environmental file under
/project/XX/physical/env dir.\n";
    print "\n";
   print " This command creates all tcl files needed for PNR under
/project/XX/physical/TCL\n";
    print "\n";
}
```

1.6 Summary

In this chapter, the concept of data structures and their corresponding views is discussed. The data structures are categorized as repository and project. As prerequisite for a unified advanced ASIC design implementation, the organization of these data structures is discussed. Included is the importance of aggressive quality control (QC) as well as a version control system (VCS) and subversion control (SVN).

In addition, data variation based on process, voltage, and temperature (PVT) and its impact on the design are discussed.

For advanced process nodes (40 nm and below) with their large amount of data files required for libraries, physical design, Static Timing Analysis (STA), gate-level simulation, and verification and to achieve clarity among these data, a concept of naming conventions based on the following criteria is introduced:

- Process node
- Transistor type: Combination of fast and slow for PMOS and NMOS
- Parasitic extraction: Combination of minimum and maximum for capacitance and resistance
- Operating temperature: Minimum and maximum
- Operating voltage: Minimum and maximum

Bibliography

1. K. Briney, *Data Management for Researchers* (Pelagic Publishing, Exeter, 2015)
2. W. Nagel, *Subversion Version Control: Using Subversion Control System in Development Projects* (Prentice Hall PTR, Upper Saddle River, 2005)
3. K. Golshan, *Physical Design Essentials, an ASIC Design Implementation Perspective* (Springer Business Media, 2007) New York, USA

Chapter 2
Multi-mode Multi-corner Analysis

The secret to multitasking is that it isn't actually multitasking.
It's just extreme focus and organization.
Joss Whedon

Today's ASIC designs need to operate in different modes and corners. In general, an advanced ASIC design implementation flow must address different design modes such as functional modes, test modes, and several input/output functions at different PVT corners as its main function (multiple supply operating voltages).

These design modes need to be analyzed and optimized with their specific design constraints at different corners to ensure the final ASIC product performs as intended under different environmental conditions. In the end, an ASIC design must satisfy all modes and corners' timing requirements. This is known as the signoff process.

For process nodes of 45 nm and below, the process of timing signoff becomes increasingly complex due to the many corners of timing closures. Timing analysis for larger process nodes, such as 180 nm down to 45 nm, was concerned mostly with the operation for the following conditions, assuming 1.0 V as nominal operating voltage:

Best-case:

- Process: Fast PMOS and NMOS transistors
- Parasitic: Minimum resistance and capacitance
- Voltage: Best voltage (e.g., 1.1 V)
- Temperature: Best temperature (e.g., 40 °C)

Worst-case:

- Process: Slow PMOS and NMOS transistors
- Parasitic: Maximum resistance and capacitance
- Voltage: Worst voltage (e.g., 0.9 V)
- Temperature: Worst temperature (e.g., 125 °C)

© Springer Nature Switzerland AG 2020
K. Golshan, *The Art of Timing Closure*,
https://doi.org/10.1007/978-3-030-49636-4_2

However, as semiconductor manufacturers started offering different options such as multiple transistors' threshold voltage (low, standard, and high), smaller adjacent routing tracks, temperature inversion effects, and leakage concerns, the landscape of timing closures started to change. These changes in the silicon process have added the need for additional analyses.

What started as timing analysis of two operational corners (best-case and worst-case) increased to at least three additional corners. The analysis of the additional corners was needed in order to include the effect of low temperature at high voltage which impacted the timing profile of PMOS and NMOS transistors with different threshold voltages.

With advancement in silicon processing, the distance between adjacent routing tracks was reduced. In addition, the interconnect layers' width became smaller. Both of these changes introduced another corner of timing concern for capacitance coupling in that it was originally marginalized by ground and pin capacitance associated with cross-coupling and noise effect.

At 45 nm, process variation for interconnect layers added again to the number of process corners that had to be considered when timing a design. Now at 20 nm and below, the multiple PVT variations for PMOS/NMOS transistors and interconnect layers and the increased number of functional and test modes have increased the number of mode corner combinations into hundreds. For today's advanced ASIC design, multi-mode multi-corner (MMMC) timing analysis is the preferred method. MMMC timing analysis addresses several nodes and corners timing requirements and is able to analyze and optimize different timing modes and multiple corners simultaneously rather than sequentially.

2.1   Typical ASIC Design Implementation Flow

Typical ASIC design implementation consists of synthesizing the RTL design for functional mode using the worst-case corners aiming to meet the design setup timing requirement. Upon completion of design synthesis, the synthesis engineer provides the resulted netlist (pre-layout netlist) to the physical designer for place and route.

Once the pre-layout netlist is available, the physical designer starts physical design activities using worst-case mode and corner. However, during the physical design activities, the goal is to first make sure the routed design meets the physical design rules (e.g., no opened and/or shorted nets). Upon completion of the final routed design, the physical designer provides the routed or post-routed netlist to the timing analysis engineer for analysis. The timing analysis engineer examines the operation of the design across a range of process, voltage, and temperature conditions.

In typical ASIC design implementation, it is acceptable to expect that the operational space of a design is bounded by analyzing the design timing requirements at two different points. The first point is chosen by taking the worst-case condition for all three operating condition parameters (process, voltage, and temperature) for setup timing, and the second point is chosen by taking the best-case conditions for hold timing analysis for the same three parameters.

During typical ASIC design timing closures, operating nodes are divided into isolated entities and assigned to multiple timing analysis engineers. For example, the basic ASIC operating nodes and their corresponding corners are assigned as follows:

- Functional setup timing (worst-case)
- Functional hold timing (best-case)
- Scan capture hold timing (best-case)
- Scan shift hold timing (best-case)

In a typical ASIC design implementation, as shown in Fig. 2.1, the process would start with closing functional setup timing. In order to close setup timing, several

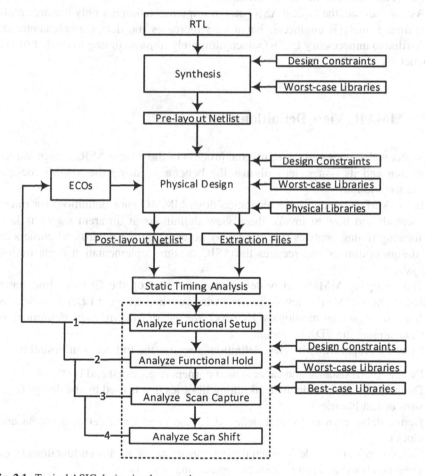

Fig. 2.1 Typical ASIC design implementation

engineering change orders (ECOs) would be required. The reason is because during the physical design stage, in order to meet ASIC, design setup timing is secondary and not a primary objective.

Once functional setup timing is met (step 1), the timing engineer begins to close functional hold timing. This leads to more ECOs for both functional setup and hold timing fixes. Fixing functional hold timing could violate functional setup timing; and fixing functional setup timing could violate functional hold timing again.

The same problem exists for scan capture hold timing fix (step 3). Fixing timing issues for scan capture hold timing has a direct effect on functional setup and hold. As a result, the process demonstrated previously is repeated.

This back and forth can be repeated several times.

Once the timing has been fixed for functional setup, functional hold, and scan capture, scan shift (step 4) hold violation needs to be fixed. Because this mode is independent of the other design modes, to fix its hold violation, the timing engineer simply adds a delay to the scan shift input.

As one can see, the typical ASIC design implementation not only requires multiple timing analysis engineers, but it also increases the design implementation cycle (due to unnecessary ECOs) which ultimately impacts timing to market of the product.

2.2   MMMC View Definitions

Now that we've reviewed the sequential process of the typical ASIC design implementation and its issues, let's discuss the benefits of using the MMMC design implementation flow.

In the MMMC design implementation flow, MMMC view definitions (or rules) are central, and how to invoke these view definitions at different stages is key. Performing timing analysis simultaneously across different modes and corners of the design optimizes and reduces the ASIC design implementation cycle timing analysis.

The following MMMC view definitions are based on the Cadence Encounter systems. These MMMC view definitions vary between different EDA providers as each use their own terminologies. However, the use of MMMC view definitions is constant among the EDA providers.

Creating proper MMMC view definitions requires the following information:

- Define RC (resistance and capacitance) corners (e.g., worst and best).
- Define a library set to include all timing libraries that is used by the design (e.g., slow or fast libraries).
- Define delay corners for cell delay (data and clock) and net delay (data and clock).
- Define constraint mode for each design constraint for all design functions (e.g., setup and hold) and test (e.g., scan capture and scan shift).

- Define analysis views for each constraint mode and delay corner.
- Define analysis views set for setup and hold for all design functions.

The following are examples of MMMC view definitions:

- Worst RC corner view at temperature 125 °C:

```
create_rc_corner -name rc_max -T 125
-qx_tech_file /common/tools/external/qrc/worst/techfiles\
-preRoute_res 1.00
-preRoute_cap 1.00
-POSTRoute_res 1.00
-POSTRoute_cap 1.00
-POSTRoute_clkres 1.00
-POSTRoute_clkcap 1.00
-POSTRoute_excap 1.00
```

- Best RC corner view at temperature −40 °C:

```
create_rc_corner -name rc_max -T -40
-qx_tech_file /common/tools/external/qrc/best/techfiles\
-preRoute_res 1.00
-preRoute_cap 1.00
-POSTRoute_res 1.00
-POSTRoute_cap 1.00
-POSTRoute_clkres 1.00
-POSTRoute_clkcap 1.00
-POSTRoute_excap 1.00
```

The routing scale factors of 1.0 for both normal nets and clock nets are default. At 1.0, there is no deviation from actual parasitic extraction. Therefore, it is highly recommended to keep the default values.

Once parasitic extractions are defined, the timing libraries used during design implementation need to be defined. Both fast (PMOS/NMOS transistors) and slow (PMOS/NMOS transistors) corners for all timing libraries that are used in the design such as standard cells, memories, IP blocks, etc. will need to be defined.

For advanced nodes, all variations of both PMOS and NMOS transistors may be included. The following MMMC view definitions are based on the fast (PMOS/NMOS transistors), slow (PMOS/NMOS transistors), worst (Cmax/Rmax), and best (Cmin/Rmin).

- Define timing library sets for slow/slow (ss) and fast/fast (ff):

```
create_library_set -name SS_LIBS -timing \
[list \
     /common/libraries/node20/stdcells_ss.lib
\
        /project/moonwalk/implementation\
        /physical/mem/mem_ss.lib \
        /common/IP/G/PLL/pll_ss.lib \
   ]

create_library_set -name FF_LIBS -timing)
[list \
     /common/libraries/node20/stdcells_ff.lib
\
        /project/moonwalk/implementation\
        /physical/mem/mem_ff.lib \
        /common/IP/G/PLL/IP_ff.lib \

   ]
```

- Define delay corners for both slow and fast libraries and their RC corners as well as their corresponding OCV (on-chip variation) derating factors of early and late for both cell and net delays. OCV factors shown in this example could vary for different process nodes:

```
create_delay_corner -name slow_max -library_set SS_LIBS \
                                            -rc_corner rc_max

            set active_corners [all_delay_corners]

            if {lsearch $active_corners slow_max]  !=-1  } {
                set_timing_derate -data -cell_delay -early \
                                    -delay_corner slow_max
0.97
                set_timing_derate -clock -cell_delay -early \
                                    -delay_corner slow_max
0.97
                set_timing_derate -data -cell_delay -late \
                                    -delay_corner slow_max
1.03
                set_timing_derate -clock -cell_delay -late \
                                    -delay_corner slow_max
1.03
                set_timing_derate -data -net_delay -early \
```

```
                                              -delay_corner slow_max
0.97
                set_timing_derate -clock -net_delay -early \
                                              -delay_corner slow_max
0.97
                set_timing_derate -data -net_delay -late \
                                              -delay_corner slow_max
1.03
                set_timing_derate -clock -net_delay -late \
                                              -delay_corner slow_max
1.03
              }

create_delay_corner -name fast_min -library_set FF_LIBS \
                                            -rc_corner rc_min

            set active_corners [all_delay_corners]


            if {lsearch $active_corners fast_min] !=-1} {
                set_timing_derate -data -cell_delay -early
\
                                          -delay_corner fast_min
0.95
                set_timing_derate -clock -cell_delay -early \
                                          -delay_corner fast_min
0.97
                set_timing_derate -data -cell_delay -late \
                                          -delay_corner fast_min
1.05
                set_timing_derate -clock -cell_delay -late \
                                          -delay_corner fast_min
1.05
                set_timing_derate -data -net_delay -early \
                                          -delay_corner fast_min
0.97
                set_timing_derate -clock -net_delay -early \
                                          -delay_corner
fast_min0.97)
                set_timing_derate -data -net_delay -late \
                                          -delay_corner
fast_min1.05
                set_timing_derate -clock -net_delay -late \
                                          -delay_corner fast_
min1.05 \
              }
```

- Define constraint modes for each design constraint:

```
                    create_constraint_mode -name setup_func_mode \
                                                    -sdc_files [
list \
                    /Project/Implementation/Synthesis/sdc/
functional.sdc]

                    create_constraint_mode -name hold_func_mode \
                                                    -sdc_files [
list \
                    /Project/Implementation/Synthesis/sdc/
functional.sdc]

                    create_constraint_mode -name hold_scans_mode \
                                                    -sdc_files [
list \
                    /Project/Implementation/Synthesis/sdc/
scan_shift.sdc]

                    create_constraint_mode -name hold_scanc_mode \
                                                    -sdc_files [
list \
                    /Project/Implementation/Synthesis/sdc/scan_cap-
ture.sdc]
```

- Define analysis view for all design modes (functional setup and hold and scan capture and shift):

```
    If { [lsearch [all_analysis_views] setup_func ]  == -1 } {
            create_analysis_view -name setup_func \
                                            -constraint_mode
setup_func_mode \
                                            -dealy_ corner
slow_max}

    If { [lsearch [all_analysis_views] hold_func ]  == -1 } {
            create_analysis_view -name hold_func \
                                            -constraint_mode
hold_func_mode \
                                            -dealy_ corner
fast_min}
```

```
If { [lsearch [all_analysis_views] hold_scanc ]  == -1 } {
        create_analysis_view -name hold_scanc \
                                       -constraint_mode
hold_scanc_mode \
                                  -dealy_ corner
fast_min}


    If { [lsearch [all_analysis_views] hold_scans ]  == -1 } {
        create_analysis_view -name hold_scans \
                                       -constraint_mode
hold_scans_mode \
                                  -dealy_ corner
fast_min}
```

The final MMMC view definition is a set of analysis views that need to be activated in various stages of the design implementation flow. The MMMC analysis view can be set as follows for an ASIC design implementation flow.

• Define analysis views set:

```
        set_analysis_view -setup [list setup_func setup_mbist] \
                             -hold [list hold_func hold_scanc
hold_scans]
```

For advanced nodes, the list for setup and hold may contain more functional modes and corners. It is important to note that the entry of both setup and hold lists has priorities. This means that once the MMMC analysis view is activated, the software works on the first entry in the list for both lists (setup and hold) aiming to close timing simultaneously. Once that is achieved, the MMMC analysis moves to the second entries of the setup and hold lists without breaking the timing of the first entries.

In the coming chapters, we will discuss which MMMC analysis view would be proper for a given stage of an advanced design implementation flow which would ensure reduction of the number of ECOs and improve the implementation cycle time.

2.3 MMMC ASIC Design Implementation Flow

As mentioned, once process nodes are moved to lower geometries (e.g., 65 and 45 nm), supply voltage levels are reduced to below 1.0 V temperature. This has a contrarian effect on a cell's delay.

Whereas in higher geometries (e.g., 130 nm and higher) technology temperature inversion had minimal to no effect on the cell's delay due to higher operating voltage (e.g., over 1.0 V), however, temperature inversion effects began to increase delay under lower operating voltage and various transistor threshold voltages.

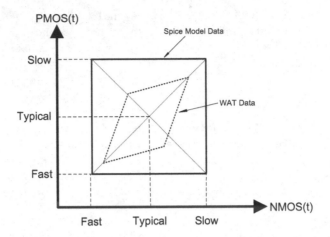

Fig. 2.2 Spice model data vs. WAT data

The direct result to timing signoff was an increase in the number of corners for signoff and optimization. The number of operating corners quickly doubled from two to four or possibly five corners for those who insisted on timing modes at typical operating conditions.

In addition, as routing pitch began to decrease, the amount of coupling capacitance increased, and that created crosstalk delay effect.

Process variations in metallization now had a non-negligible impact on the timing of the design. Plus, metal line width became small enough to impact the resistance of the wire with just a small amount of variation. Given that metallization is a separate process from base layer processing, engineers could not assume that process variation tracked in the same direction for both base and interconnect layers.

Therefore, for 45 nm and to a larger extent 20 nm node and below, multiple extraction corners were now required for timing analysis and optimization. These extraction corners consisted of worst capacitance/best resistance, best capacitance/worst resistance, and typical capacitance/resistance during physical design and timing analysis.

For higher nodes, during ASIC design synthesis, the use of slow PMOS and slow NMOS transistors was acceptable (i.e., only two corners). However, for lower process nodes, the transistor's variation (i.e., all four corners) needed to be considered during synthesis, physical design, and timing analysis.

Figure 2.2 shows an actual Spice model data for a given process versus Wafer Acceptance Test (WAT) obtained from the process control monitor (PCM).

Another phenomenon that could happen, especially during small node silicon processing, is process shift. This should not be confused with process variation.

The main problem with process shift is that yield is lost. There is no remedy to incorporate process shift during an ASIC design implementation. Rather the silicon manufacturer must extract the amount of shift and its impact on the process node performance and apply the process shift to the transistor models. Figure 2.3 shows an example of process shift.

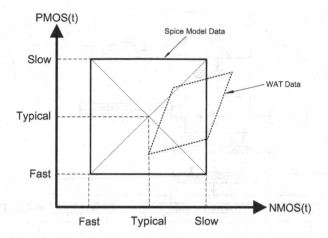

Fig. 2.3 Example of process shift

Trying to close timing on a single functional mode or process variation for a design with many using a typical design implementation introduces a critical path in another mode. MMMC will address this sequential design implementation flow through simultaneous analysis and optimization of different design timing modes.

MMMC flow in conjunction with view definition for:

- Synthesis
- Physical design
- STA

Synthesis and physical design are the areas that are most benefitted by using MMMC flow, especially when there are various design modes and multiple corners. This is because the ASIC design implementation cycle is greatly reduced due to simultaneous processing.

During synthesis, multi-mode (including variations of both PMOS and NMOS transistors) is used, but multi-corner is not. However, during physical design and STA, both multi-mode and multi-corner are used for meeting design timing constraints (setup and hold) across all design modes and corners.

The following are the multi-mode steps during synthesis:

- Read target libraries.
- Read design's RTL files.
- Elaborate design.
- Read MMMC view definitions.
- Verify constraints.
- Synthesis design.
- Perform timing analysis (design setup timing only).
- Analyze results (if timing is not met, modify design constraints or RTL).
- Output pre-layout netlist (provided timing constraints are met).

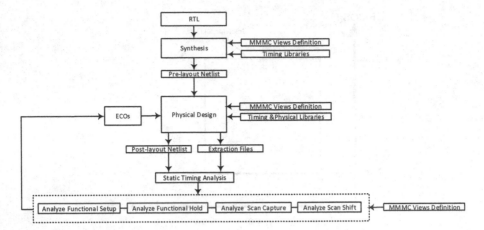

Fig. 2.4 Multi-mode multi-corner flow

Figure 2.4 shows the general MMMC flow.

Beginning with Chap. 4 and onward, how to invoke MMMC during physical design (floorplan, placement, Clock Tree Synthesis, and final route) across all the design modes and corners in order meet design timing constraints will be discussed.

2.4 Summary

Chapter 2 introduces multi-mode multi-corner (MMMC) timing analysis for an advanced ASIC design implementation flow. It explains the importance of using MMMC in contrast to a typical ASIC design implementation flow for process nodes below 40 nm.

This chapter also describes how the use of MMMC reduces the design implementation flow cycle time by minimizing the number of engineering change orders (ECOs). It also provides a clear description of how to develop MMMC views or rule definitions that would be used during design (RTL) synthesis, physical design, and Static Timing Analysis (STA).

References

1. K. Golshan, *Physical Design Essentials, an ASIC Design Implementation Perspective* (Springer Business Media, 2007) New York, USA
2. Cadence Design System Inc., *Setting Constraints and Performing Timing Analysis Using Encounter RTL Compiler* (April 2015)
3. Cadence Design System Inc., Rapid Adoption Kit, MMMC Sign of ECO Using STA & EDI System (June 2013)

Chapter 3
Design Constraints

Architecture is the triumph of human imagination over materials, methods, and men, to put man into possession of his own Earth. It is at least the geometric pattern of things, of life, of the human and social world. It is at best that magic framework of reality that we sometimes touch upon when we use the word order.
Frank Lloyd Wright

Design constraints are ASIC design specifications that are applied during synthesis (RTL to netlist), physical design, as well as STA. Each EDA tool attempts to meet these design constraints during the design's implementation process. These design constraints could be categorized as:

- Timing constraints
- Optimization constraints
- Design rule constraints

ASIC designers use an industry standard format of Synopsys Design Constraints (SDC) to define timing, optimization, and design rule constraints. These constraints are essential to meet the ASIC design's goal in terms of area, timing, and power in order to obtain the best possible implementation of the ASIC design.

The purpose of these design constraints is to ensure the design is functional and it will behave correctly after manufacturing under various PVT conditions.

3.1 Timing Constraints

The ASIC designer creates timing constraints for synthesis, physical design, and STA. These are a series of constraints applied to a given set of paths, or nets, that dictate the desired performance of a design. The major timing constraints are:

© Springer Nature Switzerland AG 2020
K. Golshan, *The Art of Timing Closure*,
https://doi.org/10.1007/978-3-030-49636-4_3

- Clocks definition (period, frequency, or speed)
- Generated clock
- Virtual clock
- Clocks skew or uncertainty
- Multi-cycle path
- Case analysis
- False paths
- Input and output delays
- Minimum and maximum path delays
- Disable timing arc

SDC format is based on the *Tcl* format, and all commands follow the *Tcl* syntax. For timing constraints, the commands are related to timing specifications of the design which contains:

- Clock definition: `create_clock`
- Generated clock: `create_generated_clock`
- Virtual clock: `create_clock`
- Clock transition: `set_clock_transition`
- Clock uncertainty: `set_clock_uncertainty`
- Clock latency: `set_clock_latency`
- Propagated clock: `set_propagated_clock`
- Disable timing arc: `set_disable_timing`
- False path: `set_false_path`
- Input and output delay: `set_input_delay` and `set_output_delay`
- Minimum and maximum delay: `set_min_delay` and `set_max_delay`
- Multicycle path: `set_multicycle_path`

Clocks need to be defined with their source (i.e., port, pin, net, or virtual) and associated characteristics (i.e., period, duty cycles, skews, and rise and fall times).

Here is an example for defining clock source from output ports of a PLL (e.g., PPLO output port/pin).

```
// clock A 10ns with 50% duty cycle
create_clock -period 10 -name CIKA -waveform {0 5} [get_ports PLLO]
```

Often, in an ASIC design, there is one or more internally generated clock sources such as clock dividers. In this case, one clock (CLKA) is external (chip input or a PLL output inside of the chip), and the other one (CLKB) is generated internally.

As shown in Fig. 3.1, INST1 provides internal clock CLKB to INST2.

Assuming clock CLKA is created from PLLO port of PLL with a period of 10 ns and symmetrical waveform (50% duty cycle), then clock CLKB is considered the generated clock at Q port of INST1. The generated clock command is as follows:

```
// Generated clock A 20ns with 50% duty cycle
create_generated_clock -divide_by 2 -source CLKA -name CLKB \
                        - waveform {3 13} [get_pins
INST1/Q]
```

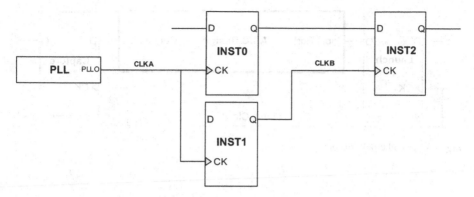

Fig. 3.1 Internally generated clock example

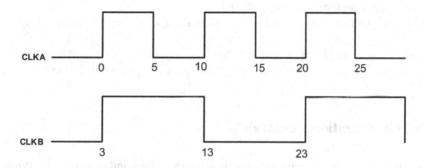

Fig. 3.2 Waveform presentation between CLKA and CLKB

The corresponding waveforms between the master clock (CLKA) and its generated clock (CLKB) are shown in Fig. 3.2.

Another clock definition is virtual clock. The virtual clock can be defined as a clock without any source. In other words, a virtual clock is a clock that has been defined, but has not been associated with any pin and port. The virtual clock does not physically exist in ASIC design. It is used as an external timing constraint for ASIC design interfaces by relating its inputs and outputs.

The virtual clock can be defined by the *create_clock* command without giving any generation point since there is no actual clock source in the design.

There are nonexistent launching and capturing registers that are used to time the design's input and output delays with a virtual clock. The launching register represents the maximum external delay for design inputs for setup timing. The capturing register represents the minimum delay for design outputs for hold timing.

The design input and outputs delays with respect to virtual clocks are only valid after Clock Tree Synthesis. Figure 3.3 illustrates virtual clock concepts.

The commands to create a design's virtual clocks are as follows:

```
//A virtual clock with 10ps period and 50% duty cycle
create_clock -name VIRTUAL_CLK -period 10 -waveform {0 5}
```

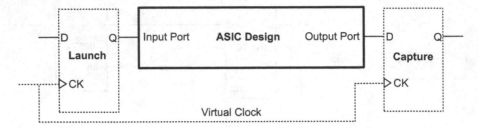

Fig. 3.3 Virtual clock concepts

```
//Set input maximum delay of 4ns
set_input_delay -clock VIRTUAL_CLK-max 4 [get_ports Input Port]

//Set output minimum delay of 2ns
set_output_delay -clock VIRTUAL_CLK -min 2 [get_ports Output Port]
```

3.2 Optimization Constraints

Optimization constraints are used to optimize speed, area, and power during synthesis and physical design.

System clocks, and their delays and maximum area, are extremely important optimization constraints in ASIC design. System clocks are typically supplied externally, but they could be generated internally for a given ASIC design. All delays such as inputs and outputs, especially in a synchronous ASIC design, are dependent upon the system clocks.

It is important to consider the relationship of an ASIC design's speed versus its area while setting these design optimization constraints. These constraints are applied by a means of cost functions. These cost functions are:

- Maximum timing delay cost
- Minimum timing delay cost
- Minimum power cost
- Maximum area cost
- Minimum area cost

Figures 3.4a, b show the relationships between an area and operational speed for a given ASIC design with respect to timing and area cost functions. As shown in these graphs, higher operational speed or frequency cost function results in larger device area (it should be noted that the maximum delay cost function has the highest priority in the cost function calculation during synthesis).

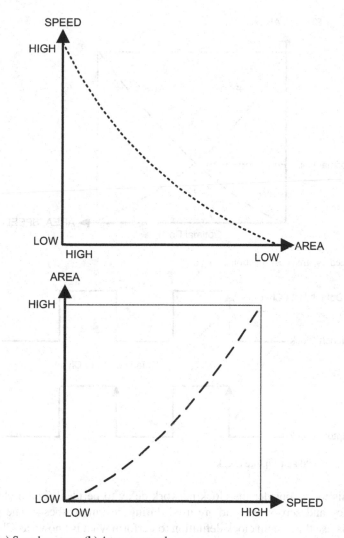

Fig. 3.4 (a) Speed vs. area. (b) Area vs. speed

Superimposing Fig. 3.4a, b (as shown in Fig. 3.5) suggests a balance between maximum delay and maximum area cost functions as an optimal cost functions setting. This is similar to defining a product price based on its supply and demand curve in economic terms.

The speed and area constraints are specified by the user. However, one should not be tempted to specify optimization constraints for maximum speed with minimum area. Doing so would cause the synthesis to run for an extensive amount of time, and if results are produced, they would be incorrect.

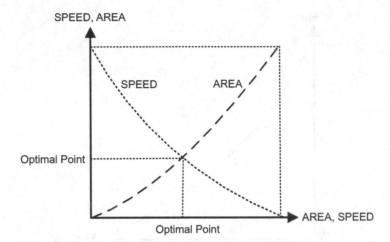

Fig. 3.5 Speed vs. area cost functions

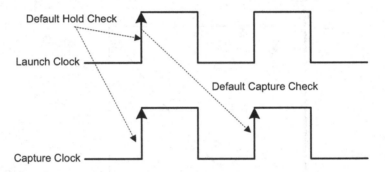

Fig. 3.6 Default hold and capture check

Synthesis tools consider the clock network delay to be ideal (i.e., a clock with fixed latency and zero skew) and are used during design synthesis. The physical design tools use the system clock definition to perform what is known as Clock Tree Synthesis (CTS) and try to meet the clock networks' delay constraints.

Generally, the combinational data path between two flip-flops (launching flip-flop and capturing flip-flop) takes a single clock cycle to propagate the data through their associated combinational logic. For high-speed ASIC design, a single clock cycle is desirable. However, in some cases, it could take more than a single clock cycle to propagate the data through the combinational data path. This is known as a multi-cycle path.

Using a multi-cycle path constraint, one could define the data as being captured by a captured flip-flop. The required capture edge would then only occur after the specified number of clock cycles. If a multi-cycle path is not defined, the setup

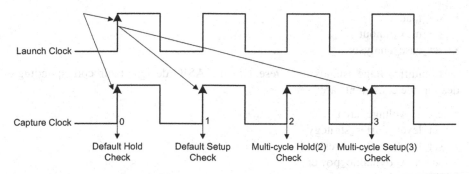

Fig. 3.7 Multi-cycle paths for setup and hold

check will occur after one clock cycle, and hold check will occur at the same edge of the captured flip-flop. This would cause a setup timing violation.

It should be noted that by default, hold check occurs at the edge of clock prior to the capture or setup check as shown in Fig. 3.6.

Assuming the combinational data path is such that it takes three clock cycles to propagate data to the capture flip-flop, then there should be a multi-cycle path defined for the setup check between launch and capture flip-flops. Then constraints for the multi-cycle path for setup and hold would be as follows (see also Fig. 3.7):

```
// multi-cycle from setup check for launching to capturing clock
Set_multi_cycle_path -setup 3 -from [get_pins launching_flop/out-
put] \
                                          -to [get_pins captur-
ing_flop/input]

// multi-cycle from hold check for launching to capturing clock
Set_multi_cycle_path -hold 2 -from [get_pins launching_flop/output]
\
                                          -to [get_pins captur-
ing_flop/input]
```

3.3 Design Rule Constraints

Design rule constraints are used to setup the environment under analysis and physical implementation of ASIC designs. The most basic design rule constraints are:

- set_driving_cell
- set_input_transition

- set_load
- set_max_fanout
- set_case_analysis

If multi-voltage islands are present in the ASIC design, their corresponding design rule constraints are:

- create_voltage_area
- set_level_shifter_strategy
- set_level_shifter_threshold
- set_max_dynamic_power
- set_max_leakage_power

Input and output delays are used to constrain the boundary of external paths in an ASIC design. These constraints specify point-to-point delays from external inputs to the first registers and from registers to the outputs.

Minimum and maximum path delays provide greater flexibility for physical design tools that have a point-to-point optimization capability. This means that one can specify timing constraints from one specific point (i.e., pin or port) in the ASIC design to another, provided such a path exists between the two specified points.

Input transition and output capacitance loads are used to constrain the input slew rate and output capacitance of an ASIC device input and output pins. These constraints have a direct effect on the final ASIC design timing.

The values of these constraints are set to zero during physical design and place and route activity to ensure that the actual ASIC design timing is calculated independent of external conditions and to make sure register-to-register timing is met. Once that is achieved, these external conditions can be applied to the design for input and output timing optimization.

False paths are used to specify point-to-point noncritical timing, either internal or external, to an ASIC design. Properly identifying these noncritical timing paths has a considerable impact on physical design tools' performance.

Design rule constraints are imposed upon ASIC design by requirements specified in each standard cell library or within physical design tools.

Design rule constraints have precedence over timing constraints because they must be met in order to realize a functional ASIC design. There are four types of major design rule constraints:

- Maximum number of fanouts
- Maximum transitions
- Maximum capacitance
- Maximum wire length

Maximum number of fanouts specifies the number of destinations that one cell can connect to each standard cell in the library. This constraint can also be applied at the ASIC design level during the physical synthesis to control the number of connections one cell can make.

Maximum transition constraint is the maximum allowable input transitions for each individual cell in the standard cell library. This constraint can also be applied to a specific net or to an entire ASIC design.

Maximum capacitance constraint behaves similarly to maximum transition constraint, but the cost function is based on the total capacitance that a standard cell can drive for any interconnection in the ASIC design. It should be noted that this constraint is fully independent of maximum transitions and, therefore, can be used in conjunction with maximum transition.

Maximum wire length constraint is useful for controlling the length of wire to reduce the possibility of two parallel long wires of the same type. Parallel long wires of the same type may have a negative impact on the noise injection and could cause crosstalk.

These design rule constraints are mainly achieved by properly inserting buffers at various stages of physical design. Thus, it is imperative to control the buffering during place and route in order to minimize area impact.

3.4   Summary

Chapter 3 provides an overview of design constraints that are used during ASIC design implementation. These constraints are categorized by their functions during synthesis, physical design, and STA.

The types of constraints are timing, optimization, and design rules.

Timing constraints cover clock definitions such as system clocks, internally generated clocks, virtual clocks, multi-cycle paths, and false paths.

Optimization constraints cover speed, area, and power during synthesis and physical design.

Design rule constraints are imposed upon ASIC design by requirements specified in each standard cell library or within physical design tools.

In addition, the importance of accurate ASIC design implementation and its impact in the final (manufactured) product is discussed.

Bibliography

1. P. Kurup, T. Abbasi, *Logic Synthesis Using Synopsys* (Kluwer Academic Publishers, Dordrecht, 1995)
2. K. Golshan, *Physical Design Essentials, an ASIC Design Implementation Perspective* (Springer Business Media, 2007) New York, USA

Chapter 4
Floorplan and Timing

A designer has a duty to create timeless design. To be timeless you have to think really far into the future, not next year, not in two years but in 20 years minimum.
Philippe Starck

Floorplanning is the art of any physical design. A well-thought-out floorplan leads to an ASIC design with higher performance and optimum area.

Floorplanning can be challenging in that it deals with the placement of I/O pads and macros as well as power and ground structures.

Before one proceeds with the physical floorplan, one needs to make sure that the data used during physical design activity is prepared properly. Proper data preparation is essential to all ASIC physical design in order to implement a correct-by-construction design. This is especially important when dealing with multi-mode and multi-corner.

Entire physical design phases may be viewed as transformations of the representation in various steps. In each step, a new representation of an ASIC design is created and analyzed. These physical design steps are iteratively improved to meet system requirements. For example, the placement or routing steps are iteratively improved to meet the design timing specifications.

Another challenge commonly faced by physical designers is the occurrence of timing and/or design rule violations during final ASIC design timing analysis and/or verification (logical and physical). If such violations are detected, the physical design steps need to be repeated to correct the errors. These error corrections have a direct impact on the ASIC cycle time because it may require the entire physical design steps be repeated in order to meet timing specifications and/or to correct verification errors.

Most of the time, these corrections are very time-consuming. Therefore, one of the objectives of the physical designer is to reduce the number of iterations during each step of the design with respect to timing and verification.

The first step in any physical design is to use high-quality and well-prepared data such as libraries (logical and physical), technology files (e.g., LEF), design constraints, incoming pre-layout netlists (i.e., timing), and operational scripts.

© Springer Nature Switzerland AG 2020
K. Golshan, *The Art of Timing Closure*,
https://doi.org/10.1007/978-3-030-49636-4_4

This becomes more important when dealing with advanced nodes (e.g., 40 nm and below) and MMMC style. Therefore, it's imperative to do a check and balance during the floorplan stage, which is one of the earliest steps in physical design.

Assuming for a given process node all logical and physical libraries have gone through an extensive QC flow, the first step for floorplanning is to modify configurational, general setting and floorplan scripts of the physical design tools to reflect project requirements. Sample scripts are shown under the **Floorplan Scripts** section in this chapter.

4.1 Floorplanning Styles

Efficient design implementation of any ASIC requires an appropriate style or planning approach that enhances the implementation cycle time and allows the design goals, such as area and performance, to be met. There are two style alternatives for design implementation – flat and hierarchical. For small to medium ASICs, flattening the design is most suited; for very large and/or concurrent ASIC design, partitioning the design into sub-modules, or hierarchical style, is preferred.

The flat implementation style provides better area usage but requires effort during physical design and timing closure compared to the hierarchical style. This area advantage is mainly due to there being no need to reserve extra space around each sub-design partition for power, ground, and resources for the routing.

Timing analysis efficiencies arise from the fact that the entire design can be analyzed at once rather than analyzing each sub-circuit separately and then analyzing the assembled design later. The disadvantage of this method is that it requires a large memory space for data and run time increases rapidly with design size.

The hierarchical implementation style is mostly used for very large and/or concurrent ASIC designs where there is a need for a substantial amount of computing capability for data processing. In addition, it is used when sub-circuits are designed individually.

However, hierarchical design implementation may degrade the performance of the final ASIC. This performance degradation is mainly because the components forming the critical path may reside in different partitions within the design, thereby extending the length of the critical path.

Therefore, when using a hierarchical design implementation style, one needs to assign the critical components to the same partition or generate proper timing constraints in order to keep the critical timing components close to each other, thereby minimizing the length of the critical path within the ASIC.

In the hierarchical design implementation style, an ASIC design can be partitioned logically or physically.

Logical partitioning takes place in the early stages of ASIC design (i.e., RTL coding). The design is partitioned according to its logical functions, as well as physical constraints, such as interconnectivity to other partitions or sub-circuits within the design. In logical partitioning, each partition is placed and routed separately and placed as a macro, or block, at the ASIC top level.

Physical partitioning is performed during the physical design activity. Once the entire ASIC design is imported into physical design tools, partitions can be created. Most often, these partitions are formed by recursively partitioning a rectangular area containing the design and using vertical or horizontal cut lines.

Physical partitioning is used for minimizing delay and satisfying timing and other design requirements in a small number of sub-circuits. Minimizing delay is subject to the constraints applied to the cluster for managing circuit complexity.

Initially, these partitions have undefined dimensions and fixed area (i.e., the total area of cells or instance added to the partition) with their associated ports, or terminals, assigned to their boundaries such that the connectivity among them is minimized. In order to place these partitions, or blocks, at the chip level, their dimensions as well as their port placement must be defined.

Perfection often requires iteration. It is recommend that once RTL design is nearing its design completion (e.g., 80% completed), the physical designer should complete the full physical design from floorplan through physical verification, power analysis (such a leakage), and STA to ensure readiness of the entire physical and timing flow.

Once the physical design database is created, the first step is to modify default values in the physical design *Tcl* scripts (as described in Chap. 1) according to design requirements (e.g., process node, libraries, etc.).

Before importing the Verilog pre-layout (synthesized) netlist, there are several types of *Tcl* scripts that need to be modified. These scripts are:

- Configuration (*moonwalk_config.tcl*)
- Design setting (*moonwalk_setting.tcl*)
- MMMC view definition (*moonwalk_view_def.tcl*)
- Power and ground (*moonwalk_png.tcl*)
- Power management (*moonwalk_cpf.tcl*)
- Design floorplan ((*moonwalk_flp.tcl*)

Configuration script is physical design tool-related and contains controlling log files, formatting timing report files, and custom productive procedures (e.g., counting the usage of standard cells with different threshold voltages, automatically collecting all short nets in the design and repairing them). In addition, if one finds a useful productive procedure during the physical design process for a given design, then it can be added to the tool's master configuration file so as to be available for future projects.

Design setting scripts are design-specific settings that apply to all stages of physical design and need to be modified according to the design and its associated standard cell libraries.

In general, these design settings establish which standard cells should not be used (e.g., very low drive standard cells), critical nets that should be buffered and/or touched (e.g., manually routed nets), and so on.

MMMC view definitions script is used for concurrent timing analysis and is physical design stage-dependent. During the floorplan stage, it is used for timing checks for all modes and corners in the design; during the placement stage, for setup timing analysis; during the Clock Tree Synthesis stage, for setup and hold timing analysis; and during the design's final route stage, for timing analysis for all modes and corners. The details of this script are discussed in Chap. 2.

Power and ground scripts are primarily used during physical design floorplanning. This script provides standard cells' power and ground mesh after the manual core and macro power and ground are constructed.

For technology nodes such as 90 nm and higher, the standard practice was to put power and ground next to each other. However, for advanced nodes, this type of placement will short the power and ground. This is due to advanced nodes having smaller metal layer spacing which causes the short between power and ground if there is particle contamination during the silicon process. To avoid this, one needs to alternate placement of power and ground.

The floorplan *Tcl* script for *Moonwalk* project includes power management sections (i.e., power management keyword was set to true in design environmental file as discussed in Chap. 1).

Before starting floorplanning, the design needs to be timed by activating the MMMC mode for all functional modes for setup timing checks. At this stage, hold timing checks are not required because all clocks are ideal.

For setup timing checks, the design must meet required timing with no or very small violations due to differences between synthesis and physical design tools of their timing engines (i.e., timer).

If large setup timing violations are observed (e.g., larger than 70 ps or so), they need to be investigated before continuing. Large setup timing violations may have several sources.

Design constraint issues (e.g., missing false paths, multi-cycle paths, and/or unrealistic external input/output delay) may be the cause of these violations. Or the design may have been synthesized using single mode rather than multi-mode. In this case, one could perform netlist-to-netlist optimization within the physical design tool option in order to optimize the incoming netlist using MMMC. This, of course, is dependent upon the tool supporting such an option. Otherwise, the MMMC method would need to be used during synthesis.

Once no major violations are found, one can proceed with floorplanning.

The first step would be to make the chip power and ground ring. Next, place hard macros, such as memories and PLL. Do not apply any power and ground strips and/ or mesh. Instead, run a coarse placement.

After coarse placement, all modules in the design should be highlighted using different colors to see which modules are clustered together. If the modules are not clustered together, it is mainly due to the placement of hard macros. They need to be placed so that module clustering occurs. Proper module clustering improves timing and minimizes routing.

However, if some of the modules are tightly clustered (i.e., area congestion is too high), one could use the area utilization option of the design tool to reduce the congestion. High congestion of the area makes the region unrouteable and could adversely impact timing.

Floorplan refining is an iterative process until the optimal floorplan is achieved. Figure 4.1 illustrates a refined floorplan after quick placement.

Often the checks that detect problems with the physical database are related to the netlist, such as unconnected ports, mismatched ports, standard cell errors, or errors in the library and technology files. The physical design tool generates a log file that contains all errors and warnings. It is important to review the log file and make sure that all reported errors and warnings are resolved before proceeding to the next phase.

With smaller process geometry (e.g., 20 nm or below), increasing speed and number of gates are going into the silicon. This is due to smaller power consumption and could be limiting for many applications.

Hence, an important aspect of today's ASIC design is to manage power and reduce its consumption.

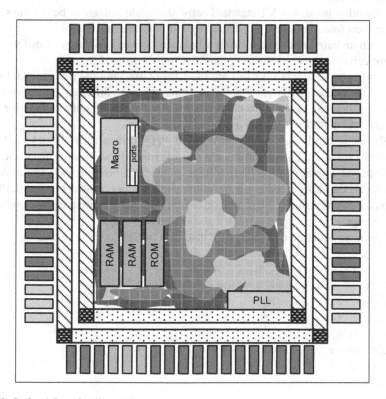

Fig. 4.1 Refined floorplan illustration

There are various techniques for reducing the power. These techniques are:

- Dynamic voltage and frequency scaling (DVFS)
- Using multiple voltage threshold (VT) standard cells
- Multiple power supply and level shifter (low and high) insertion
- Power gating (shutting off a portion of the design)

DVFS utilizes frequency-driven voltage regulator (FDVR). As the system's clock frequency increases, so does the output voltage of the regulator suppling device power. And as the system clock's frequency decreases, so does the output of the voltage regulator. Thus, the device's dynamic power is reduced.

The use of multiple VT standard cells (high, standard, and low) is accomplished during the physical design optimization process.

Low VT standard cells increase leakage, and, therefore, usage should be minimized. This is especially important on advanced nodes using an ultra-low-power process as the final ASIC performance could be adversely affected.

While high VT standard cells have the least leakage power, its voltage threshold is very close to the supply voltage, and it has the potential of having a temperature inversion effect.

On the other hand, low VT standard cells' threshold voltage is below the supply voltage; therefore, they are less affected by temperature inversion.

As demonstrated above, achieving a balance between the uses of different VT standard cells is critical in ensuring the ASIC performs as designed.

Multiple power supply domains are used for conserving the dynamic and static power of the design. Different functional domains run at different supply voltages. In this way, one can save the power losses by reducing the supply voltage for standard cells and memory elements in the design.

Different power domains are defined based on the criticality of the design. In this style of floorplanning, level shifters are used for the signal coming from the low-voltage power domain to the high-voltage power domain and vice versa. At the netlist level, the design code will be written in Cadence Power Format (CPF) or Universal Power Format (UPF) based on which one can develop the power structure for the design.

Below is an example of level shifter insertion in CPF:

```
### High-to-Low level shifters ###
define_level_shifter_cell -cells LVS-H2L* \
-input_voltage_range 0.9:1.0:1.1 \
-output_voltage_range 0.8:1.0:0.1 \
-direction down \
-output_power_pin VDD \
-ground VSS \
-valid_location to
```

```
### Low-to-High Level Shifters ###
define_level_shifter_cell -cells LVS-L2H* \
-input_voltage_range 0.8:1.0:0.1 \
-output_voltage_range 0.9:1.0:1.1 \
-input_power_pin VDD-IOW\
-output_power_pin VDD \
-direction up \
-ground VSS \
-valid_location to
```

From a floorplanning perspective, there are some special placement guidelines for inserting the level shifter across the different power domains in the design.

The level shifter should be placed in the destination domain of the design. There is one disadvantage of inserting the level shifter – it occupies area in design. But at the same time, inserting the level shifter will help in saving the power of the device.

Figure 4.2 shows high- to low-level shifter, and Fig. 4.3 low- to high-level shifter.

From a timing perspective, these level shifters have a minimal impact on timing (they are similar to a buffer). However, it's important to note that for high- to low-level shifters, the low-voltage swing input signal is not necessarily strong enough to turn the input transistor fully on. This could lead to an unacceptably long rise or fall time, which could then result in a higher switching current and reduced noise margin.

The last technique used for reducing the power is known as power gating. This is especially effective for leakage power reduction versus dynamic power reduction. In this technique, one would shut down a portion of the design by means of disconnecting their supplies either from the power or ground.

There are two types of power gating. One is fine-grain which adds power-down transistors to every standard cell used in the shutdown domain. This can create timing issues introduced by intercluster voltage variations that are difficult to resolve.

The second type of power gating is coarse-grain which implements the grid style of power-down domain and drives standard cells through a virtually powered network. This approach is less sensitive to the process variations and introduces less power leakage. In coarse-grain power gating, the power gating transistors are a part of the power distribution network rather than the standard cell.

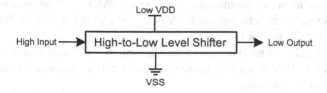

Fig. 4.2 High- to low-level shifter

Fig. 4.3 Low- to
high-level shifter

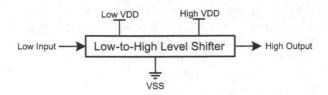

Fig. 4.4 Example of
header cell

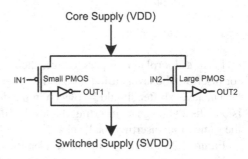

In this technique, the blocks are placed in the shutdown mode when the function is not active, and turned on when required, using two types of power switches. One uses PMOS transistors as the power-down cell. This is known as the header cell for the power supply (VDD) control.

The other method uses NMOS transistors as the ground cell. This controls the ground supply (VSS) and is known as the footer cell.

Because of the effectiveness of using the header cell over the footer cell for power gating, the header cell will be discussed.

The header cell contains two sizes of PMOS transistors: one small and one large, in regard to their gate length. Using a small PMOS transistor brings up the power-down domain (switched supply) slowly before the large PMOS transistor is turned on. This is done in order to prevent a large amount of current surge which has the potential of damaging standard cells' gates within the power-down domain during the powering up procedure.

Figure 4.4 shows a header cell circuit. In this example, the header cell has one small PMOS transistor with input (IN1) and inverted output (OUT1). The same goes for the large PMOS transistor with input (IN2) and inverted output (OUT2).

From a physical design point of view, when the header cells are placed next to each other, they form two chains—a chain of small PMOS transistors and a chain of large PMOS transistors. The last small PMOS transistor in the chain will turn on the first large PMOS transistor in its chain.

The header cells are used to build a ring around the switched power domain.

Either CPF or UPF is used to define the header cell, isolation cells, and always-on and power-down domain.

An example of the power-down domain's CPF is shown below:

```
set_cpf_version 1.0
```

```
#### Isolation Cells ###
define_isolation_cell -cells ISO_AND \
-power VDD \
-ground VSS \
-enable ISO \
-valid_location to
### Isolation Rule ###
create_isolation_rule -name ISORULE -from PWRDOWN \
-isolation_condition "!PWRDOWN/isolation_enable" \
-isolation_output high
update_isolation_rules -names ISORULE -location to -cells ISO_AND
### Power Switch cell (header) ###
define_power_switch_cell -cells {HEADER} \
-power_switchable SVDD -power VDD \
-stage_1_enable !TPWRON1 \
-stage_1_output IPWRON2 \
-stage_2_enable !PWRON2 \
-stage_2_output I ACKNOWLEDGE \
-type header
### Always-On (AO) Domain ###
create_power_domain -name AO -default
create_power_nets -nets VDD  -voltage 0.8
create_ground_nets -nets VSS
update_power_domain -name AO -internal_power_net VDD
create_global_connection -domain AO -net VDD -pins VDD
create_global_connection -domain AO -net VSS -pins VSS
### Power Down (PWRDOWN) Domain ###
create_power_domain -name core -instances PWRDWN \
-shutoff_condition {PWRON1/pwron_enable}
create_power_nets -nets SVDD -internal -voltage 0.8
create_ground_nets -nets VSS
update_power_domain -PWRDOWN -internal_power_net SVDD
create_global_connection -domain PWRDOWN -net VSS -pins VSS
create_global_connection -domain PWRDOWN -net VDD_SW -pins SVDD
create_power_switch_rule -name PWRSW -domain PWRDOWN \
-external_power_net VDD
update_power_switch_rule -name PWRSW \
-cells HEADER \
-prefix PWR_SW_ \
-acknowledge_receiver ACKNOWLEDGE
```

During floorplanning, once the power-down domain region is defined, a header cell ring is constructed around it. Figure 4.5 shows the structure of header cell's chain. Additionally, there are two cells—start and end.

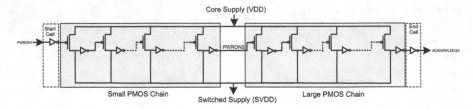

Fig. 4.5 Header ring chain structure example

With the start cell, when PWRON1 is low, the power domain is in the shutdown mode. The end cell provides a non-inverted signal from the last header cell's output. This is used as *acknowledge* (high) indicating the power domain is ON.

According to Fig. 4.5, in order to bring the power-down domain to an ON state, PWRON1 input signal needs to go from a low to high state which will turn on the small PMOS transistor chain.

Once all small PMOS transistors are turned on, its last stage triggers the first stage of the large PMOS transistor chain by the PWRON2 input going from a low to high state.

Once all large PMOS transistors are turned on, the power-down domain's power supply will be ON by connecting VDD (core supply) to SVDD (switched VDD) through the PMOS transistors' R-on resistance in the header cells.

To ensure their floating outputs of the power-down domain are not propagating into the always-on domain when the power-down domain is in shutdown mode, isolation cells are used. In this method, isolation cells are ANDed and/or NANDed with *acknowledge* with the outputs of the power-down domain preventing any floating output signals.

One needs to consider the IR drop across the header ring. Depending upon the header cell's layout, its R-on resistance could be obtained through Spice simulation. Once the header cell's R-on resistance is determined, it can be used to calculate the number of header cells in the ring. The higher the header cells count, the lower the ring IR drop.

In contrast to the chip's input/output pad ring, the header ring library contains header cells, header filler cells, start cell, end cell, power protection (CLAMP) cell, outer corner cell, and inner corner cell. Figure 4.6 shows a conceptual floorplan that has both multipower and power gating domain styles.

In the floorplan, once multipower and power gating domains are defined, level shifters can be placed manually, and the header ring can be constructed and placed properly by the physical design tools. See *moonwalk_flp.tcl.* in the **Floorplan Scripts** section.

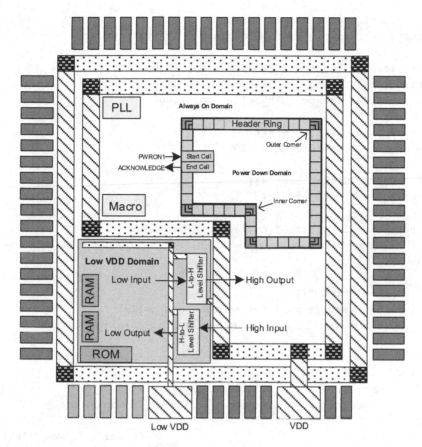

Fig. 4.6 Conceptual floorplan with different power domains

4.2 Floorplan Scripts

Place and route tool's specific configuration script

```
etMultiCpuUsage -localCpu 8


suppressMessage"LEFPARS-2036"
### Pin is not on the manufacturing grid ###
suppressMessage"ENCLF-82"
### Appending library 'USERLIB' to the previously read library of
the same name ###
suppressMessage"TECHLIB-459"
```

```
set_global report_timing_format \
{instance cell pin fanout load delay arrival required}

### Disables loading ECSM data from timing libraries ###
#set_global timing_read_library_without_ecsm true
set delaycal_use_default_delay_limit 1000

### Temporary increase ###
setMessageLimit 1000 ENCDB 2078

### Count VT Cells Usage ###
 proc multivt_counter {} {

   set count_lvt [sizeof_collection [get_cells\
        [get_cells -hierarchical -filter"is_hierarchical == false"
] \
                                        -filter"ref_name=~*LVT*
|| ref_name=~*lvt*"]]
   set count_svt [sizeof_collection [get_cells\
        [get_cells -hierarchical -filter"is_hierarchical == false"
] \
                                        -filter"ref_name=~*SVT*
|| ref_name=~*svt*"]]
  set count_hvt [sizeof_collection [get_cells\
        [get_cells -hierarchical -filter"is_hierarchical == false"
] \
                                        -filter"ref_name=~*HVT*
|| ref_name=~*hvt*"]]
   set count_all [sizeof_collection [get_cells -hierarchical -filter"
is_hierarchical == false" ]]

   set count_svt [expr { $count_all - $count_lvt - $count_hvt }]
   set percent_hvt [expr { $count_hvt * 100.00 / $count_all }]
   set percent_lvt [expr { $count_lvt * 100.00 / $count_all }]
   set percent_svt [expr { $count_svt * 100.00 / $count_all }]

   echo" ### Total Cell Count = $count_all ###"
   echo"  HVT Cell % = $percent_hvt\\%"
   echo"  LVT Cell % = $percent_lvt\\%"
   echo"  SVT Cell % = $percent_svt\\%"
   echo" end" }

 ### Remove Clock NETS from None Default Rule list to Relief
Routing congestions ###
 proc remove_ndr_nets {ndr_nets} {
```

```
   foreach net $ndr_nets {
      editDelete -net $net
      setAttribute -net $net -non_default_rule default   }   }

### Delete and Reroute Shorted NETS ###
proc get_nets_from_violations {{skip_pg_net 0} {in_file {}}} {
  set n_list {}
  set fd""
  if {[string length $in_file] > 0} { set fd [open $in_file w] }

  foreach mkr [dbGet top.markers.subType Short -p ] {
    set msg [dbGet $mkr.message]
    if {[regexp {Regular Wire of Net (S+)} $msg junk str1] || \
       [regexp {Regular Via of Net (S+)} $msg junk str1]} {
        if {$skip_pg_net == 0 || ![dbGet [dbGet top.net.name
$str1 -p].isPwrOrGnd] } {
         lappend n_list $str1
       } }

    regsub {Regular Wire of Net} $msg {} next_msg
    if {[regexp {Regular Wire of Net (S+)} $next_msg junk str2]
|| \
       [regexp {Regular Wire of Net (S+)} $next_msg junk str2]} {
        if {$skip_pg_net == 0 || ![dbGet [dbGet top.net.name
$str1 -p].isPwrOrGnd] } {
        lappend n_list $str1 } }
  }
  set n_list [lsort -unique $n_list]
  if {$fd !=""} {
    puts $fd $n_list
    close $fd }
    return $n_list
}
proc reroute_shorts {} {
  get_nets_from_violations -in_file shorts.rpt
  deselectAll
  set NETFILE [open"shorts.rpt" r]
  foreach i [read $NETFILE] { editDelete -nets $i -type Signal}
  close $NETFILE }
```

Design specific setting scripts

```
setDesignMode -process 20
#set_option liberty_always_use_nldm true
```

```
set_interactive_constraint_modes [all_constraint_modes -active]


### Do Not Use these Libraries and Cells ###
set_dont_touch [get_cells -hier *spare*] true
set_dont_touch [get_cells -hier *SCAN_dt*] true
set_dont_touch [get_cells -hier SCAN_dt*] true

foreach   cell   [list       {node20_*/*d0p5}        {node_20_*/*d0}
{node_20_*/*dly*}   ] { \
  setDontUse $cell true }

foreach net [list  VDD_MEM  XTALI  XTALO ] {\
  set_dont_touch [get_nets $net] true
  setAttribute -net $net -skip_routing true }

foreach no_buffer_net [list critical nets ] } \
  set_dont_touch [get_nets $no_buffer_net] true }
```

Design standard cell power and ground strips, mesh, and rings
The coordinates (e.g., X1, Y1, X2, and Y2) for creating these power and ground nets are obtained for the floorplan.

```
### Create Core Power and Ground Ring ###
setAddRingMode -avoid_short 0 -extend_over_row 1
addRing -nets {VDD VDD_MEM VSS} -width 10 -offset 5 -spacing 3 \
        -layer {top M5 left TM2 bottom M5 right TM2}  -type core_
rings \
        -follow io

### VDD_MEM Area Only ###
addStripe -nets {VDD VDD_MEM VSS} -area {X1 Y1 X2 Y2} \
          -direction vertical -layer M4 -width 1.5 -spacing 16.5\
          -set_to_set_distance 54 -orthogonal_only 0 \
          -padcore_ring_bottom_layer_limit M3 \
          -padcore_ring_top_layer_limit M5 \
          -block_ring_bottom_layer_limit M3 \
          -block_ring_top_layer_limit M5

addStripe -nets {VDD VDD_MEM VSS} -area {X1 Y1 X2 Y2} \
          -direction horizontal -layer M5 -width 1.5 -spacing 16.5
\
```

```
                -set_to_set_distance 54 -orthogonal_only 0 \
                -padcore_ring_bottom_layer_limit M4 \
                -padcore_ring_top_layer_limit TM2 \
                -block_ring_bottom_layer_limit M4 \
                -block_ring_top_layer_limit TM2


### Always ON Domain Area Only ###
addStripe -nets {VDD VSS} -area_blockage {X1 Y1 X2 Y2} \
                -direction vertical -layer M4 -width 1.5 -spacing 16.5 \
                -set_to_set_distance 36  -orthogonal_only 0 \
                -padcore_ring_bottom_layer_limit M3 \
                -padcore_ring_top_layer_limit M5 \
                -block_ring_bottom_layer_limit M3 \
                -block_ring_top_layer_limit M5


addStripe -nets {VDD VSS} \
                -area_blockage {X1 Y1 X2 Y2} \
                    -direction vertical -layer TM2 -width 6 -spacing 30
-set_to_set_distance 72 \
                -orthogonal_only 0 \
                 -padcore_ring_bottom_layer_limit M5 -padcore_ring_top_
layer_limit TM2 \
                -block_ring_bottom_layer_limit M5 -block_ring_top_layer_
limit TM2 \
                -start_x 16.75

### Power Down Area Only ###
addStripe -nets {VSS SVDD} -direction vertical \
                -layer M4 -width 1.5 -spacing 16.5  -set_to_set_distance
36 \
                -orthogonal_only 0 -xleft_offset 0 -over_power_domain 1
\
                -padcore_ring_bottom_layer_limit M3  -padcore_ring_top_
layer_limit M5 \
                -block_ring_bottom_layer_limit M3  -block_ring_top_layer_
limit M5


addStripe -nets {VSS SVDD} -direction horizontal \
                -layer M5 -width 1.5 -spacing 16.5 -set_to_set_distance
36 \
                  -orthogonal_only 0 -ybottom_offset 12.6 -over_power_
domain 1 \
                -padcore_ring_bottom_layer_limit M4  -padcore_ring_top_
layer_limit TM2 \
                -block_ring_bottom_layer_limit M4  -block_ring_top_layer_
```

```
limit TM2

### Adding Redistribution Layer (RDL) ###
addStripe -nets {VDD VSS} -direction horizontal \
          -layer RDL -width 6.0 -spacing 12 -set_to_set_distance
36 \
          -orthogonal_only 1 -stacked_via_top_layer RDL -stacked_
via_bottom_layer TM2 \
          -padcore_ring_bottom_layer_limit TM2  -padcore_ring_top_
layer_limit RDL \
          -block_ring_bottom_layer_limit TM2   -block_ring_top_
layer_limit RDL

sroute -connect floatingStripe -nets VSS
editSelect -net VSS
editMerge

### Adding Standard Cell Power/Ground Rails ###
sroute -connect {corePin} -layerChangeRange {M1 M4} \
       -targetViaLayerRange {M1 M4}  -deleteExistingRoutes \
       -checkAlignedSecondaryPin 1 \
       -allowJogging 0 \
       -allowLayerChange 0
```

Design floorplan script

In this script, all dimensions (e.g., X… and Y…) are obtained from the physical floorplan.

```
### Setup Design Configuration Environment ###
source    /project/moonwalk/implementation/physical/TCL/moonwalk_
config.tcl
foreach dir [list ../rpt ../RPT/flp ../RPT/plc ../RPT/cts ../RPT/
frt ../RPT/scans ../RPT/scanc ../MOONWALK ../logs] { if { ! [file
isdirectory $dir] } {exec mkdir $dir } }

### Initialize Design ###
set init_layout_view ""
set init_oa_view ""
set init_oa_lib ""
set init_abstract_view ""
set init_oa_cell ""

set init_gnd_net {VSS VSSO}
set init_pwr_net {VDD VDD_MEM SVDD  VDDO }
```

```
set init_lef_file " \
        /project/moonwalk/implementation/physical/LEF/node20_PWR_
cells.lef \
    /project/moonwalk/implementation/physical/LEF/clock_NDR.lef \
    /project/moonwalk/implementatio/physical/LEF/node20_stdcells_
hvt_pg.lef \
    /project/moonwalk/implementatio/physical/LEF/node20_stdcells_
svt_pg.lef \
    /project/moonwalk/implementatio/physical/LEF/node20_stdcells_
svt_pg.lef \
        /project/moonwalk/implementation/physical/LEF/node20_io_
pad_35u.lef \
    /project/moonwalk/implementation/physical/LEF/LOGO.lef \
    /project/moonwalk/implementation/physical/LEF/PLL.lef \
    /project/moonwalk/implementation/physical/LEF/ROM.lef \
    /project/moonwalk/implementation/physical/LEF/RAM.lef \
set init_assign_buffer "1"

set init_mmmc_file "/project/moonwalk/implementatio/physical/TCL/
moonwalk_view_def.tcl"
set init_top_cell moonwalk
set  init_verilog  /project/moonwalk/implementatio/physical/NET/
moonwalk_pre_layout.vg

## Import Verilog Netlist ###
set_analysis_view -setup [list All Setup Modes ] -hold [list All
Hold Modes ]
init_design

globalNetConnect VDD -type pgpin -pin VDD -netlistOverride
globalNetConnect VSS -type pgpin -pin VSS -netlistOverride

globalNetConnect VDD -type TIEHIGH
globalNetConnect VSS -type TILOW

### Load and commit CPF ###
loadCPF   /project/moonwalk/implementation/physical/CPF/moonwalk.
cpf
commitCPF -verbose

### Uncomment  to  preserve  ports  Optimization  on  any  specific
module(s) ###
#set modules [get_cells -filter "is_hierarchical == true" CORE/
CLK_GEN*]
#getReport {query_objects  -limit 10000} > ../keep_ports.list
```

```
#setOptMode -keepPort ../keep_ports.list

### Uncomment for Netlist-to-Netlist optimization ###
#source    /project/moonwalk/implementatio/physical/TCL/moonwalk_
setting.tcl
#set_analysis_view -setup [list All Setup Modes ] -hold [list All
Hold Modes ]

#runN2NOpt -cwlm \
#          -cwlmLib ../wire_load/moonwalk_wlm.flat \
#          -cwlmSdc ../wire_load/moonwalk_flat.sdc \
#          -effort high \
#          -preserveHierPinsWithSDC \
#          -inDir n2n.input \
#          -outDir n2n.output \
#          -saveToDesignName n2n.enc
#freedesign
#source    /project/moonwalk/implementation/physical/TCL/moonwalk_
config.tcl
#source n2n.enc
#source    /project/moonwalk/implementatio/physical/TCL/moonwalk_
setting.tcl
### End of Netlist-to-Netlist optimization ###

deleteTieHiLo -cell TILOW
deleteTieHiLo -cell TIHIGH

source /project/moonwalk/implementatio/physical/TCL/moonwalk_set-
ting.tcl


saveDesign -tcon ../MOONWALKf/flp_int.enc -compress
timeDesign - setup - flp_int  -expandedViews -numPaths 1000 -outDir
../RPT/flp -prefix INT

## Setup Chip FloorPlan Boundary ###
floorPlan -siteOnly unit -coreMarginsBy die -d X1 Y1 X2 Y2 X3 Y3

defIn /project/moonwalk/implementation/physical/DEF/moonwalk_pad.
def


foreach side [list top left bottom right] {
```

```
   addIoFiller   -prefix pd_filler   -side $side \
       -cell {pd_rfill10 pd_rfill1 pd_rfill101 pd_rfill001 pd_rfill0005}
}


### Power Domain Section ###
setObjFPlanBoxList Group PDWN_CORE X1 Y1 X2 Y2 X3 Y3
modifyPowerDomainAttr PDWN_CORE -minGaps 28 28 28 28
modifyPowerDomainAttr PDWN_CORE -rsExts 30 30 30 30
dbSelectObj [dbGet -p2 top.fplan.rows.site.name ao_unit]
deleteSelectedFromFPlan
initCoreRow -powerDomain PDWN_CORE


## Add Power Switch Ring ###
addPowerSwitch -ring \
     -powerDomain PDWN_CORE   -power {(VDD:VDD)(SVDD:SVDD)} \
     -ground {(VSS:VSS)} \
     -enablePinIn {PWRON2} -enablePinOut {PWRONACK2 } \
      -enableNetIn {i_alwayson/pmu_ctrl_16} -enableNetOut {core_
swack_1 } \
     -specifySides {1 1 1 1 1 1 1 1 1 1} -sideOffsetList {3 3 3 3
3 3 3 3 3 3} \
     -globalSwitchCellName {{HEADER S} {HEADER_CLAMP D}} \
      -bottomOrientation MY -leftOrientation MX90 \
      -topOrientation MX -rightOrientation MY90 \
     -cornerCellList HEADER_OUTER \
     -cornerOrientationList {MX90 R180 MX90 MX MY MX MY90 R0 MY90
MY} \
     -globalFillerCellName {{HEADER_FILLER}} \
      -insideCornerCellList HEADER_INNER \
     -instancePrefix SW_CORE_ \
     -globalPattern {D S S S S S S S S S} \
     -continuePattern


set   CORE_SWITCH   [addPowerSwitch   -ring   -powerDomain   PDWN_
CORE -getSwitchInstances]



rechainPowerSwitch -enablePinIn {PWRON2} -enablePinOut {PWRONACK2}
\
       -enableNetIn {i_alwayson/pmu_ctrl_16} -enableNetOut {core_
swack_1} \
     -chainByInstances -switchInstances $CORE_SWITCH


rechainPowerSwitch -enablePinIn {PWRON1} -enablePinOut {PWRONACK1}
\
```

```
      -enableNetIn {core_swack_1} -enableNetOut {core_switch_ack_
out} \
      -chainByInstances -switchInstances $CORE_SWITCH

### Add Instance Dose Not in The Netlist ###
addInst LOGO LOGO

setInstancePlacementStatus -all HardMacros -status fixed


### Creating a  Soft Power Region ###
createInstGroup uc -isPhyHier
addInstToInstGroup uc i_alwayson/uc/*
createRegion uc 1940 1830 2780 3115

### For Magnet Instance Placement ###
#place_connected -attractor hard_macro -attractor_pin CLK -level
1 -placed

source    /project/moonwalk/implementation/physical/TCL/moonwalk_
png.tcl


defOut  -floorplan  /project/moonwalk/implementation/physical/def/
moonwalk_flp.def


### To be Used for Incremental Design Synthesis ###
#defIn /project/moonwalk/implementation/physical/def/moonwalk_flp.
def

timeDesign -prePlace -expandedViews -numPaths 1000 -outDir ../RPT/
flp
timeDesign -prePlace -hold -expandedViews -numPaths 1000 -outDir
../RPT/flp
summaryReport -noHtml -outfile ../RPT/flp/flp_summaryReport.rpt
saveDesign -tcon ../MOONWALK/flp.enc -compress

if { [info exists env(FE_EXIT)] && (FE_EXIT) == 1 } {
  exit
}
```

4.3 Summary

Chapter 4 explains various basic aspects of physical design data preparation and ASIC physical design and floorplanning alternatives.

In the area of data preparation, several timing and design constraints, and their impact on the quality of the ASIC physical design floorplan, are discussed.

Also, the fundamentals of different floorplanning styles, mainly flat and hierarchical, and their advantages are addressed. Likewise, a review of basic floorplanning techniques is offered.

Construction of power domains, either multi-voltage or power gating, is discussed. Level shifters (low to high and high to low) are used in the construction of a multi-voltage domain.

For the power gating style, header cells are used to construct the header ring. The header ring includes the header, header filler, outer corner, inner corner, and power protection (CLAMP) cells. In conjunction, a sample of Cadence Power Format (CPF) rule descriptions is included.

In this chapter, a sample of creating physical design and floorplanning scripts is provided with respects to how to time the incoming netlist quality using multi-mode multi-corner (MMMC) and the usage of netlist optimization.

It should be noted that floorplanning style depends upon many factors, such as the type of ASIC, area, and performance, and relies heavily on one's physical design experience.

Bibliography

1. N. Sherwani, *Algorithms for VLSI Physical Design Automation*, 2nd edn. (Kluwer Academic Publishers, Dordrecht, 1997)
2. K. Golshan, *Physical Design Essentials, an ASIC Design Implementation Perspective* (Springer Business Media, 2007) New York, USA
3. D. Chinnery, K. Keutzer, *Closing the Power Gap between ASIC & Custom: Tools and Techniques for Low Power Design* (Springer Science + Business Media, 2007) New York, USA
4. Cadence Design Systems, Inc., *Rapid adoption kits based on a 10nm reference flow for new arm cortex* (2016)

Chapter 5
Placement and Timing

For a house to be successful, the objects in it must communicate with one another, respond and balance one another.
Frank Lloyd Wright

The goal of standard cell placement is to map ASIC components, or cells, onto positions of the ASIC core area (i.e., standard cell placement region), which is defined by rows. The standard cells must be placed in the assigned region (i.e., rows) such that the ASIC can be routed efficiently and the overall timing requirements can be satisfied. Standard cell placement of an ASIC physical design has always been a key factor for achieving physical design with optimized area usage, routing congestion, and timing requirements.

Almost all of today's physical design tools use various algorithms to place standard cells automatically. Although these placement algorithms are very complex and are being improved frequently, the basic idea has remained the same. There are several key factors that need to be considered in preventing or minimizing timing violations that may occur during the later stages of physical design, such as Clock Tree Synthesis (CTS), final route, and STA.

Placement takes two steps. One is coarse placement which is connectivity-driven. The objective is to cluster cells within a given module in the design near each other (i.e., clustering). The second step is fine placement and optimization which is timing-driven. Its objective is to meet design setup timing requirements for all functional modes based on the MMMC definitions. Since the clocks are ideal during this stage, there will be no hold timing optimization.

Before coarse placement (during importing the netlist at floorplan or at placement), the buffering and inverter pairs will be removed and replaced with new buffers. In general, to save area, the new buffers will be selected from the smallest standard cells (i.e., cells with lowest drive strength) in the library.

Because of this, these buffers will introduce timing violations later, and the only way to up-size them is through the ECO process. From a timing perspective, in order to avoid usage of the lowest drive strength cells from the library (i.e., any cell with less than a factor of 10 drive strength), the physical designer assigns a "do not

© Springer Nature Switzerland AG 2020
K. Golshan, *The Art of Timing Closure*,
https://doi.org/10.1007/978-3-030-49636-4_5

use" attribute. It is important that this attribute be removed before starting CTS and final route so as to allow EDA tools to use these cells for timing violation correction.

5.1   TAP and Endcap Cells

There are two types of standard cell design. One is to have standard cells with an internal tap. In other words, the N-well is tied to the power supply (VDD), and the substrate is tied to the ground supply (VSS) as shown in Fig. 5.1.

Another type of standard cell design is when standard cells do not have their own tap (tap-less), and so are using external tap cells for N-well and substrate connections to VDD and VSS as shown in Fig. 5.2.

Having external tap cells allows standard cell designers to reduce the area of standard cells by sharing multiple standard cells with one tap cell.

In larger technology nodes, having N-well and substrate tied to VDD and VSS within standard cells was not an important area benefit due to the silicon processing design rules. For example, contact to diffusion spacing rules was of little consequence in comparison with other design rules such as routing layers.

However, for advanced nodes, the processing design rules are shrinking, and by using tap-less standard cell design allows for a smaller area. To put it into perspective, in larger technology nodes, *contact* programmable read-only memory (ROM) was smaller than *via* programmable ones in area. For advanced node technology, it is other way around.

Fig. 5.1 Standard cell with internal tap

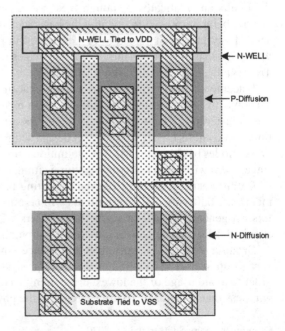

Fig. 5.2 Tap-less standard cell

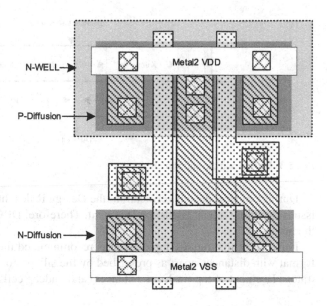

It is important to note that since the internal taps are removed from standard cells, their PMOS and NMOS transistor's source and drains must be connected to VDD and VSS. This is because *metal1* is no longer available to make these connections. Therefore, the only option is to use the *metal2* layer (Fig. 5.2) in a horizontal direction similar to VDD and VSS of standard cells' supply rails. This changes the overall design's routing layer directions.

For non-tap-less design, the routing scheme was odd horizontal metal layers and even vertical layers. However, for tap-less design, *metal2* is the same as *metal1* for horizontal layers in order to connect power/ground supply rails to the standard cells. Having *metal2* layers in a horizontal direction has an advantage in that one can increase its width for improving the design's IR drop—provided there are no routing congestions.

The importance of using tap and endcap cells is to increase resistance between power and ground connections for N-well and substrate separately. This avoids a phenomenon known as *latch-up*.

Latch-up is a short circuit between power and ground which is caused by a low-resistance path between power and ground rails that are subject to an interaction between parasitic PNP and NPN bipolar transistors. This interaction is inherent in the CMOS process. The high current produced by this phenomenon damages PMOS and NMOS transistors.

The rules for taps and endcaps are very technologically dependent. Some technologies don't require them at all because the tap connections are built-in to their standard cells (generally for larger technology nodes). For advanced technology nodes that are using tap-less standard cells, a tap cell is required to be placed evenly as prescribed by the silicon foundries with the endcap cells placed at the end of every standard cell row as shown in Fig. 5.3.

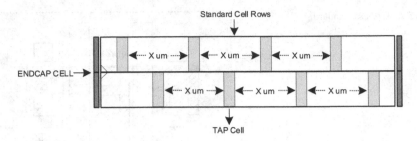

Fig. 5.3 Endcap and tap cell placement

During physical design verification, the Design Rule Check (DRC) will flag any issues regarding tap and endcap placement. Therefore, DRC should be run early in the process.

For designs that require tap cells, it is recommended they be placed in a matrix format with distances apart as prescribed by the silicon foundries. Figure 5.3 illustrates placement of tap (in Xum distances) and endcap cells.

5.2 Spare Cell Insertion

Although spare cells are not required, they can provide insurance so that if bugs are discovered after the design is manufactured, the design can be corrected at minimal cost.

Spare cells allow one to go back later and do a small functional correction ECO to the design. The spare cell could be used from standard cell libraries and/or from special libraries called metal programmable cell ECO libraries.

There are two ways of inserting spare cells. One is known as a shotgun approach which places spare cells randomly in the design area. The second way is to create predefined spare cell clusters from standard cells and/or metal programmable cell ECO libraries and place them in an orderly manner in the design area as shown in Fig. 5.4.

It should be noted that spare cell clusters can be inserted in a design that has multiple power domains as well.

5.3 High Fanout Nets

The placement algorithm is programmed to fix design rule constraint requirements such as max transition, max capacitance, and high fanout nets.

Max transitions could be in the standard cell library for each cell or could be defined by the user. Max transition is used to limit long signal transition. Long signal transition reduces the cell's noise margin, increases net delay, and could cause a

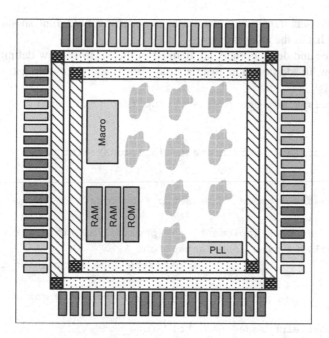

Fig. 5.4 Conceptual spare cell cluster insertion

short circuit (PMOS and NMOS transistors to be on at the same time) which increases leakage current.

Max capacitance and max fanout values are mainly user-defined. They are used to reduce gate output capacitance loading for a given net (excluding clock nets) which, in turn, drives many input gates.

Although max transition, max capacitance, and max fanout can be eliminated during synthesis, it is not recommended. Rather, they should be eliminated during design placement.

During design placement, max transition, max capacitance, and max fanout are fixed by inserting what is known as a buffer tree. At this stage, buffer tree insertions are not as structured as the ones used for CTS.

Although the user can set max transition value in the physical design tool, if the libraries have a value for max transition, the value in the libraries has precedence over that which was user-defined.

It is important to check both best-case and worst-case libraries for their max transition values. If the values are the same, no action is required.

However, if best-case and worst-case libraries have different max transition values, the user needs to include both libraries during max transition fix.

By default, only the worst-case library is used during placement which means only those max transition violations in the worst-case library will be fixed. Because of this, the best-case library max transition violations would not be fixed and would appear during hold timing analysis in the signoff phase. Fixing these violations during that stage will require many ECOs in order to correct.

   To prevent max transition violations from occurring, hold_func should be added
to the setup list of the analysis view.
   All modes and delay corners are defined in the MMMC view definition file as
shown below. In addition, note that the hold_func is added to setup list:

```
set_analysis_view -setup [list setup_func hold_func]


set_interactive_constraint_modes [all_constraint_modes -active]


update_constraint_mode -name setup_func_mode \
-sdc_files [list \
 /project/moonwalk/implementation/physical/SDC/moonwalk.sdc]
cerate_analysis_view -name setup_func \
                              -constraint_mode -name setup_func_mode
\
                              -delay_corner slow_max

update_constraint_mode -name hold_func_mode \
-sdc_files [list \
 /project/moonwalk/implementation/physical/SDC/moonwalk.sdc]
cerate_analysis_view -name hold_func \
                              -constraint_mode -name hold_func_mode
\
                              -delay_corner fast_min
```

   By adding hold_func to the setup list, the physical design tool will be aware of
both worst-case and best-case libraries and their max transition requirements. Buffer
trees will be inserted based on tiered requirements.
   Once the max transitions from both worst-case and best-case libraries are met,
the hold_func needs to be removed from the setup list of the analysis view.

```
set_analysis_view -setup [list setup_func] -hold [ list hold_func]
```

   As mentioned before, max fanout and max capacitance values are used to reduce
the gate output capacitance loading which drives many input gates. Examples of
input gates may include reset, scan enable, and scan clock.
   During the placement process while using the functional setup constraint, scan
clock is attributed as a high fanout net in the design. Therefore, the physical design
tool will apply a max fanout fix with no conditions, as it does not differentiate
between scan clock and other high fanout nets.
   This becomes a timing problem during the final timing analysis of scan capture
hold timing violations. Correcting scan capture hold timing violations will often
introduce new setup timing violations and may require many ECOs to correct both

the functional setup and scan capture hold timing. This occurs because the functional setup timing mode and scan capture mode are using two different design constraints.

Functional design constraints often have several clocks with multiple clock frequencies. However, the scan constraints have only one clock (scan clock) with only one frequency. This difference becomes problematic during the functional setup timing and scan capture hold timing processes.

In a typical physical design flow, users concentrated on the functional setup and hold timing issues. Later they would manually correct scan capture hold timing violations through the ECO process. Using MMMC flow will eliminate or minimize these types of ECO.

In order to take full advantage of MMMC during signoff, one needs to identify the scan clock net by finding its source and using the "do not touch" attribute. This should be done during the placement phase and before proceeding with the high fanout fix.

Once high fanout fix is completed for all noncritical nets in the design (e.g., reset net), "do not touch" attributes need to be removed from the scan clock net and proceed with the high fanout fix for the scan clock using the balanced buffer insertion option (most physical design tools support this option) as shown in Fig. 5.5.

The advantage of inserting balanced buffers on the scan clock net during placement is to provide the ability of reducing scan clock skews by adjusting the inserted balance buffers at input of scan MUXs with minimal manual ECOs and without distributing functional clock tree structures in the design. This is shown in Fig. 5.6.

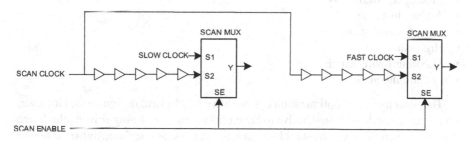

Fig. 5.5 Scan clock balanced buffer insertion

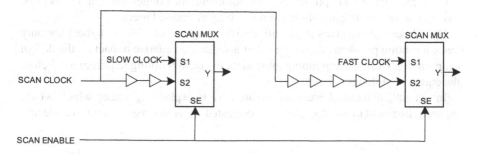

Fig. 5.6 Minimizing scan clock skew buffer adjustment

5.4 Placement

Depending upon the physical design tool, placement algorithms have many options. Choosing the correct placement option leads to meeting the design's timing requirements as well as reduction in area.

Design placement is a two-step process—placement of standard cells (i.e., placement mode) and optimization of those placements (i.e., placement optimization mode) according to timing constraint requirements.

Placement mode options may be as follows:

- Wire length reduction
- Uniform density
- Maximum density
- Congestion reduction
- Timing-driven
- Clustering
- Clock gate awareness
- Clock tree insertion awareness
- Maximum clock tree clustering fanout

Placement optimization mode options may be as follows:

- Optimization effort
- Leakage power reduction
- Dynamic power reduction
- Clock gate awareness
- Adding instances
- Use of useful skew
- High fanout fix
- Maximum wire length
- Reclaiming area

The placement and optimization options vary by physical design tools. However, for advanced nodes, it's imperative to have options such as timing-driven, clock tree insertion awareness, clustering, clock gate awareness, and leakage/dynamic power reduction.

Once placement and optimization are complete, the congestion map needs to be reviewed in order to ensure there are no heavily congested areas.

Heavily congested areas (e.g., with utilization factor of 97% or higher) not only create a routing problem (shored nets) but also have a negative impact on the design setup timing. Design setup timing must meet the design timing requirements before placement completion.

In general, congested areas are created by floorplanning issues which would require refinement of the floorplan. Or congested areas may be inherent to a specific module in the design which has a very high back-to-back connectivity amount in its standard cells.

An example of high back-to-back connectivity is a synthesized small-sized ROM. While this is not a recommended practice, some RTL designers have the misconception that having a synthesized small ROM saves area (which is true at the netlist level). However, this causes routing and timing issues during placement and final routing. One remedy to resolve this type of congestion problem would be to create a region for the module with high back-to-back connectivity with a lower utilization factor in the floorplan.

After placement, there should not be any large setup timing violations (e.g., over 100 ps). If there are, they need to be investigated as to their root cause and then corrected.

The sources of setup timing violations could be floorplan and/or design constraints, such as missing case analysis, multi-cycle paths, false paths, over-constrained input/output delays, and/or clock gating cells that are not real clock gating cells (in that case, they would need to be disabled).

5.5   Placement Scripts

The following placement script shows spare cell module insertion for power-down and low-power domains. The script also includes standard cell placement and setup timing optimization.

```
### Setup Environment ###
source    /project/moonwalk/implementation/physical/TCL/moonwalk_
config.tcl

### Placement Settings ###
restoreDesign ../MOONWALK/flp.enc.dat moonwalk
source    /project/moonwalk/implementation/physical/TCL/moonwalk_
setting.tcl

generateVias


set_interactive_constraint_modes [all_constraint_modes -active]
set_max_fanout 50 [current_design]
set_max_capacitance 0.300 [current_design]
set_max_transition 0.300 [current_design]

update_constraint_mode -name setup_func_mode \
```

```
   -sdc_files [list /project/moonwalk/implementation/physical/SDC/
moonwalk_func.sdc]
create_analysis_view -name setup_func \
   -constraint_mode setup_func_mode  -delay_corner slow_max

update_constraint_mode -name hold_func_mode \
   -sdc_files [list /project/moonwalk/implementation/physical/SDC/
moonwalk_func.sdc]
create_analysis_view -name hold_func \
   -constraint_mode hold_func_mode  -delay_corner fast_min

set_interactive_constraint_modes [all_constraint_modes -active]
source    /project/moonwalk/implementation/physical/TCL/moonwalk_
clk_gate_disable.tcl

set_analysis_view -setup [list setup_func] -hold [list hold_func]


setPlaceMode \
   -wireLenOptEffort medium \
   -uniformDensity true \
   -maxDensity -1 \
   -placeIoPins false \
   -congEffort auto \
   -reorderScan true \
   -timingDriven true \
   -clusterMode true \
   -clkGateAware true \
   -fp false \
   -ignoreScan false \
   -groupFlopToGate auto \
   -groupFlopToGateHalfPerim 20 \
   -groupFlopToGateMaxFanout 20

setTrialRouteMode -maxRouteLayer 12


setOptMode \
   -effort high \
   -preserveAssertions false \
   -leakagePowerEffort none \
   -dynamicPowerEffort none \
   -clkGateAware true \
   -addInst true \
   -allEndPoints true \
```

```
     -usefulSkew false \
     -addInstancePrefix IPO_ \
     -fixFanoutLoad true \
     -maxLength 600   -reclaimArea true
```

Do not optimize the module ports that is used for Gate Level
Simulation ###
setOptMode -keepPort ../keep_ports.list

```
createClockTreeSpec \
    -bufferList {   hvt_ckinv_lvtd24   hvt_ckinv_lvtd16   hvt_ckinv_
lvtd12 hvt_ckinv_lvtd10 \
                        hvt_ckbuffd24 hvt_ckbuffd16 hvt_ckbuffd12 }
-file moonwalk_plc.spec
```

```
cleanupSpecifyClockTree
specifyClockTree -file moonwalk_plc.spec
```

Spare Cell Modules Placement

Place All Non-Spare Instances at Location (0,0) With Status
FIXED ###


```
set nonSpareList [dbGet [dbGet -p [dbGet -p -v top.insts.isSpare-
Gate 1].pstatus unplaced].name]
foreach i $nonSpareList {placeInstance $i 0 0 -fixed}
```

```
placeDesign
```


```
dbDeleteTrialRoute
```


```
createSpareModule \
  -moduleName SPARE \
  -cell {hvt_sdfsrd4 hvt_nd2d4 hvt_aoi22d4 hvt_buffd24 hvt_invd12
}   -useCellAsPrefix
```

```
placeSpareModule  -moduleName SPARE -prefix SPARE -stepx 400 -stepy
400 -util 0.8
```

```
placeSpareModule -moduleName SPARE -prefix SPARE -stepx 400 -stepy
400 \
   -powerDomain PDWN_CORE -util 0.8

## Fix Spare Cells Placement ###
set spareList [dbGet [dbGet -p top.insts.isSpareGate 1].name]
foreach i $spareList {dbSet [dbGet -p top.insts.name $i].pstatus
fixed}

### Unplace Non-Spare Cells ###
foreach i $nonSpareList {dbSet [dbGet -p top.insts.name $i].psta-
tus unplaced}

### Placement ###


### Don't Use LVT and SVT Cells for Placement ###
foreach cell [list {lvt*/*} ]
    { setDontUse $cell true }
foreach cell [list {svt*/*} ]
    { setDontUse $cell true }

placeDesign -inPlaceOpt
timeDesign -prects -prefix IPO -expandedViews -numPaths 1000 -out-
Dir ../RPT/plc

saveDesign ../MOONWALK/ipo.enc -compress


### Pre-CTS Placement Optimization ###


set_interactive_constraint_modes [all_constraint_modes -active]
setOptMode -addInstancePrefix PRE_CTS_
setCTSMode -clusterMaxFanout 20

optDesign -preCTS


clearDrc


verifyConnectivity \
    -type special \
    -noAntenna \
```

```
  -nets { VSS VDD } \
  -report ../RPT/plc/plc_connectivity.rpt

verifyGeometry \
  -allowPadFillerCellsOverlap \
  -allowRoutingBlkgPinOverlap \
  -allowRoutingCellBlkgOverlap \
  -error 1000 \
  -report ../RPT/plc/plc_geometry.rpt

clearDrc


setPlaceMode -clkGateAware false
setOptMode -clkGateAware false

timeDesign -preCTS  -prefix PLC -expandedViews -numPaths 1000 -out-
Dir ../RPT/plc
saveDesign -tcon ../MOONWALK/plc.enc -compress

summaryReport -noHtml -outfile ../RPT/plc/plc_summary_report.rpt


### Provide Project Specific Wire Load Model for Synthesis ###
#wireload -outfile ../wire_load/moonwalk_wlm -percent 1.0 -cell-
Limit 100000

if { [info exists env(FE_EXIT)] && $env(FE_EXIT) == 1 } {exit}
```

5.6 Summary

Chapter 5 discusses the aspects of advanced node physical design placement and how to resolve design timing violations at the placement stage. In addition, the different standard cell designs (with internal tap and external tap) are explained as well as how to place tap cells according to the silicon foundries avoiding violations of their design rules.

In this chapter, the importance of balanced routing of the scan clock in order to help scan capture timing closure with minimal ECO is discussed. Also, two phases of placement are explained—standard cell placement and their timing optimization.

Bibliography

1. J.E. Vilson, J.J. Liou, Electrostatic Discharge in Semiconductor Devices: An Overview. Proc. IEEE **86**(2), 399–420 (1998)
2. S.A. Campbell, *The Science and Engineering of Microelectronic Fabrication,* (Oxford University Press, 2007) Oxford UK
3. K. Golshan, *Physical Design Essentials, an ASIC Design Implementation Perspective* (Springer Business Media, 2007) New York, USA

Chapter 6
Clock Tree Synthesis

The only reason for time is so that everything doesn't happen at once.
Albert Einstein

The concept of Clock Tree Synthesis (CTS) is the automatic insertion of buffers/inverters along the clock paths of the ASIC design in order to balance the clock delay to all clock inputs.

In order to balance clock skew and minimize insertion delay, CTS is performed. Naturally, before CTS, all clock pins are driven by a single clock source and considered as an ideal net.

A typical ASIC design could contain many clock sources with different frequencies. This has made use of CTS challenging. Without efficient clock gating and clock tree implementation, design timing and power reduction cannot be achieved.

Before CTS, it's important to understand the design's clock structures and balancing requirements in order to have proper exceptions and to be able to construct optimal clock trees.

Some prerequisites for CTS are:

- Creating non-default rules (e.g., setting large metal width and spacing for clock routing)
- Setting clock's max transition, max capacitance, and max fanout
- Selecting which cells (clock buffer, clock inverter) to use during CTS (although clock buffers have equal rise and fall time, to avoid pulse width violations, it is recommended to use clock inverters that balance both rise and fall time simultaneously)
- Setting CTS exceptions

As mentioned before, CTS plays an important role in building well-balanced clock trees, fixing timing violations, and reducing the extra unnecessary pessimism in the design. The goal during building a clock tree is to reduce the skew, to maintain symmetrical clock tree structure, and to cover all the registers in the design and reduce power consumption.

© Springer Nature Switzerland AG 2020
K. Golshan, *The Art of Timing Closure*,
https://doi.org/10.1007/978-3-030-49636-4_6

To have well-balanced clock trees, one must understand the design clocks' latency and skew requirements provided by design constraints. However, using design constraints during CTS may cause unnecessary clock cell insertion and create timing violations. Typically, these issues arrive from clock dividers (i.e., generated clocks), unconstrained low and fast clock multiplexers (i.e., missing case analysis), default clock gating cells (i.e., non-clock gate cells), or cross-domain clocks (e.g., slow clock logically *or/and* with fast clock) .

To avoid unnecessary clock cell insertions and meet design timing and skew requirements, three stages of CTS are recommended. These stages are:

- Building physical clock tree structure with CTS constraints
- Optimizing clock tree with CTS constraints
- Final clock tree timing optimization with actual design constraints

6.1 Building Physical Clock Tree Structures

Building physical clock tree structures is the first stage (CTS1) of CTS. The objectives at this stage are to build a physically well-balanced clock tree, to avoid excessive clock cell (clock buffer and/or clock inverter) insertion, and to make sure the clock skew is as minimal as possible. In addition, all design timing requirements must be met.

The first step is to have CTS constraints (e.g., *moonwalk_cts.sdc*) which are created from copying the actual design's functional design constraints (e.g., *moonwalk_func.sdc*) without any modification.

The second step is to add additional MMMC modes for CTS (e.g., setup_cts_ mode and hold_cts_mode) as shown here:

```
update_constraint_mode -name setup_cts_mode \
-sdc_files   [list   /project/implementation/physical/SDC/moonwalk_
cts.sdc

create_analysis_view -name setup_cts \
                            -constraint_mode setup_cts_mode \
                            -dealy_ corner slow_max

update_constraint_mode -name hold_cts_mode \
-sdc_files   [list   /project/implementation/physical/SDC/moonwalk_
cts.sdc

create_analysis_view -name hold_cts \
                            -constraint_mode hold_cts_mode \
                            -dealy_ corner fast_min
```

In addition to selecting cells (i.e., clock buffers and/or clock inverters) to be used during CTS, one needs to provide CTS options such as max cluster fanout (i.e., the number of leaf cells in the design driven by one clock tree cell), max capacitance, and max clock transition. In addition, for timing analysis, on-chip variation (OCV) and common pessimism path removal (CPPR) need to be set.

OCV concerns arise during the ASIC manufacturing process. To account for process variation, derating factors are used. Derating factors can be used for timing check (setup and hold), net delay, and cell delay in the design and have the following format:

```
set_timing_derate -early  0.9
set_timing_derate -late   1.1
```

For example, the above constraints derate the early/minimum by −10% (i.e., the path becomes faster) and latc/maximum by +10% (i.e., the path becomes slower). For setup timing check, launch clock paths (i.e., late/maximum) will be multiplied by -late option, and capture clock (i.e., early/minimum) will be multiplied by -early option.

For hold timing check, the clock and launch paths (i.e. early/minimum) will be multiplied by -early option, and captured clock (i.e. late/maximum) will be multiplied by -late option.

It is important to note that for advanced nodes, using OCV is pessimistic because of the fixed derating factors which were applied to the entire chip. A more realistic application of derating factors for such nodes is advanced OCV (AOCV). In contrast to OCV, AOCV derating factors increase as the distance increases and is, therefore, less pessimistic. AOCV is a distance- based (global) and path-based (local) model for derating.

CPPR arises when the launching and capturing clocks share a common path. The difference between the max delay and min delay of this common clock path segment is called the common path pessimism and is removed during timing analysis.

If there are any clock cell cluster instances (e.g., back-to-back buffers and inverters), one needs to use their physical design tool debug and diagnosis viewer (e.g., clock tree browser) to understand these clustering and resolve before continuing (uniform clock tree insertion after clock gating cell (CG) is desired) (Fig. 6.1).

Figure 6.2 shows a conceptual clocking browser indicting three back-to-back clock cell cluster instances. One is for a divider flip-flop (FF), second is for unconstrained mux (MUX) between slow and fast clocks, and third is cross-domain clock gating (GATE) between slow and fast clocks.

Once excessive clock cell insertion and large setup violations are observed, one should modify the CTS constraints by adding exceptions to certain flip-flop's clock port (i.e., sink pin), converting clock definition from generated to created, disabling clock gating checks on non-clock gating cells, and/or adding more case analysis and false paths to CTS constraints in order to minimize clock cell insertions and improve clock tree structures.

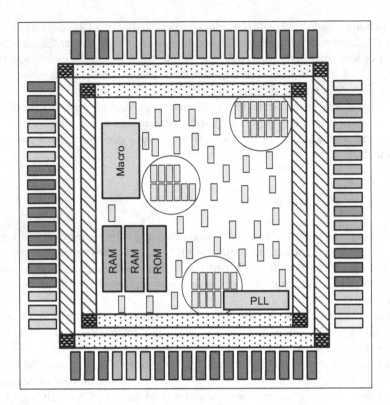

Fig. 6.1 Illustration of back-to-back clock tree insertions

Based on the applied modified CTS constraints and exceptions, all leaf cells (i.e., flip-flop) are aligned. In other words, there are minimal overall skews and no routing congestion due to back-to-back clock cell clustering.

Removal of excessive clock cells is an iterative process so that optimal clock tree structures are achieved. This is shown in Fig. 6.3.

There may some critical clock nets in the design that should not be touched by CTS. For those critical clock nets, one needs to insert buffer/inverter cells manually at the floorplan and set with a "do not touch" attribute.

For optimal propagation delay, these buffers/inverters may be selected to monotonically increase in their drive strength d by a factor of α for each level of clock tree path and set. This is shown in Fig. 6.4.

It is important to note that some physical design tools will move spare cells during CTS by default. One needs to set an option to avoid moving spare cells:

```
set_ccopt_property change_fences_to_guides false
```

To ensure clock tree max fanout (e.g., 30) during CTS, set the following option:

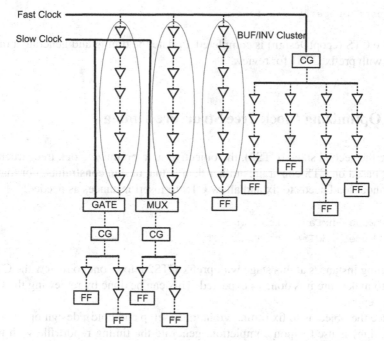

Fig. 6.2 Conceptual excessive clock cell (back-to-back) insertion

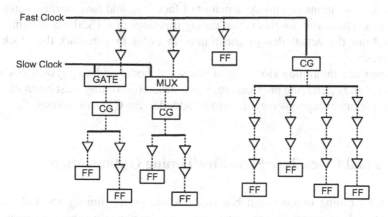

Fig. 6.3 Conceptual optimal clock tree structures

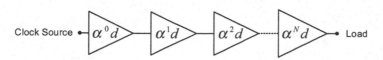

Fig. 6.4 Manual buffer/inverter insertion

```
set_ccopt_property max_fanout 30
```

Once CTS (ccoptDesign) is completed, update the timing and generate a timing report with prefix CTS1 for review.

6.2 Optimizing Clock Tree Structure Timing

During the second stage (CTS2), the objective is to optimize clock tree structures' timing based on CTS constraints rather than actual design constraints. For that, the high fanout load needs to fix and allow CTS to insert instances as needed:

```
setOptMode -fixFanoutLoad true
setOptMode -addInstancePrefix CTS2_
```

Adding instances at this stage with prefix CTS2 allows one to review the CTS in order to make sure it is done as expected. This can be done by reviewing the CTS2 timing report.

Since the object is to fix timing violations (setup and hold), design optimization (optDesign) is used. Upon completion, generate the timing report file with prefix CTS2 and review.

It is important to note that after design timing optimization, there should not be any timing violations (setup). As a matter of fact, it should have positive setup timing slacks. The reason for this is that during this stage of CTS, the constraints used depart from the actual design constraints in order to construct the clock tree structures.

If there are any timing violations at this stage, both building physical clock tree structures and optimizing physical clock tree structures' timing must be repeated. In addition, more exceptions may need to be added in the CTS constraints.

6.3 Final Clock Tree Structure Timing Optimization

Final CTS timing optimization has two options—setup timing and hold timing fixes—based on actual design constraints (e.g., *moonwalk_func.sdc*) rather than using CTS constraints (e.g., *moonwalk_cts.sdc*) that were used in previous steps.

Update MMMC modes for both setup and hold timing based on functional design constraints:

```
update_constraint_mode -name setup_func_mode \
-sdc_files   [list  /project/implementation/physical/SDC/moonwalk_
func.sdc

create_analysis_view -name setup_func \
                          -constraint_mode setup_func_mode \
                          -dealy_ corner slow_max

update_constraint_mode -name hold_func_mode \
-sdc_files   [list   /project/implementation/physical/SDC/moonwalk_
func.sdc

create_analysis_view -name hold_func \
                          -constraint_mode hold_func_mode \
                          -dealy_ corner fast_min
```

And to invoke MMMC:

```
set_analysis_view -setup [list setup_func]  -hold [list hold_func]
```

During the optimization mode, fix the clocks' max fanout by allowing clock buffer and inverter insertion:

```
setOptMode -fixFanoutLoad true
setOptMode -addInstancePrefix CTS
```

During CTS timing optimization using MMMC, setup timing violations are fixed first, and the hold timing violations are fixed if their timing violation fixes do not break the setup timing. This process will occur concurrently. Since setup timing violation fixes have priority over hold timing violation fixes (according to the analysis view setting), there may be some hold violations that need to be fixed explicitly. But before fixing, design timing needs to be updated (i.e., timeDesign) for both setup and hold timing generated for both.

To fix the remaining hold timing violations explicitly, first remove "do not use" attributes for small drive buffers and delay cells, as these were set during placement and CTS:

```
setDontUse dly1d1 false
setDontUse dly2d1 false
setDontUse dly3d1 false
setDontUse bufdp5 false
setDontUse bufd2p5 false
```

Hold optimization modes need to be set, which includes not modifying the clock structure by honoring clock domains. In addition, do not allow setup timing to be violated, and do not fix any register-to-output and input-to-register hold timing violations. In addition, provide a list of buffer and delay cells which will be used during hold timing violation fixes:

```
setAnalysisMode -honorClockDomains true


setOptMode -fixHoldAllowSetupTnsDegrade false
setOptMode -ignorePathGroupsForHold { reg2out in2reg }

setOptMode -holdFixingCells { \
                    bufd2p5 bufdp5 bufd2 bufd3 bufd4 dlyd1 dlyd2
dlyd3 }

setOptMode -addInstancePrefix  HLD_FIX
```

Optimize the design with "hold" option (optDesign -postCTS -hold).
Update design timing and generate timing reports for both setup and hold timing.

6.4 Clock Tree Synthesis Script

Clock Tree Synthesis script:

```
### CTS Environment Configuration Settings ###


source    /project/moonwalk/implementation/physical/TCL/moonwalk_
config.tcl


restoreDesign ../MOONWALK/plc.enc.dat moonwalk


### Building Clock Tree Structures (CTS) with CTS Constraints  ###


### Turn on LVT for CTS (For High Speed Design) ###
#foreach cell [list {stdcells_lvt*/*} ] { setDontUse $cell false }

### Turn off HVT for CTS ###
foreach cell [list {stdcells_hvt*/* } ] { setDontUse $cell true}
```

```
source     /project/moonwalk/implementation/physical/TCL/moonwalk_
setting.tcl


generateVias


set_interactive_constraint_modes [all_constraint_modes -active]
update_constraint_mode -name setup_cts_mode \
   -sdc_files [list /project/moonwalk/implementation/physical/SDC/
moonwalk_cts.sdc]
create_analysis_view -name setup_cts \
   -constraint_mode setup_cts_mode \
   -delay_corner slow_max

update_constraint_mode -name hold_cts_mode \
   -sdc_files [list /project/moonwalk/implementation/physical/SDC/
moonwalk_cts.sdc]
create_analysis_view -name hold_cts \
   -constraint_mode hold_cts_mode \
   -delay_corner fast_min

set_analysis_view -setup [list setup_cts] -hold [list hold_cts]


set_interactive_constraint_modes [all_constraint_modes -active]
source /project/moonwalk/implementation/physical/TCL/\
moonwalk_clk_gate_disable.tcl

createClockTreeSpec \
   -bufferList { ckinvlvtd24 ckinvlvtd16 ckinvlvtd12 ckinvlvtd10 \
                 ckbuflvtd24 ckbuflvtd16 ckbuflvtd12 ckbuffd10 }
-file moonwalk_clock.spec

cleanupSpecifyClockTree
specifyClockTree -file moonwalk_clock.spec

set_interactive_constraint_modes [all_constraint_modes -active]
set_max_fanout 50 [current_design]
set_max_capacitance 0.300 [current_design]
set_clock_transition 0.300 [all_clocks]

setAnalysisMode -analysisType onChipVariation -cppr both
```

```
setNanoRouteMode \
  -routeWithLithoDriven false \
  -routeBottomRoutingLayer 1 \
  -routeTopRoutingLayer 6

setCTSMode \
  -clusterMaxFanout 20 \
  -routeClkNet true \
  -rcCorrelationAutoMode true \
  -routeNonDefaultRule rule_2w2s \
  -useLibMaxCap false \
  -useLibMaxFanout false

set_ccopt_mode \
  -cts_inverter_cells { ckinvlvtd24 ckinvlvtd16 ckinvlvtd12 ckin-
vlvtd10} \
  -cts_buffer_cells { ckbuflvtd24 ckbuflvtd16 ckbuflvtd12 ckbuffd10}
\
  -cts_use_inverters true \
  -cts_target_skew 0.20 \
  -integration native

### Keep Module Ports During CTS for Gate Level Simulation ###


#set modules [get_cells -filter "is_hierarchical == true" CORE/
clkgen_inst/*]
#getReport {query_objects $modules -limit 10000} > ../keep_ports.
list
#setOptMode -keepPort ../keep_ports.list

set_interactive_constraint_modes [all_constraint_modes -active]
set_propagated_clock [all_clocks]

set restore [get_global timing_defer_mmmc_object_updates]
set_global timing_defer_mmmc_object_updates true
set_analysis_view -update_timing
set_global timing_defer_mmmc_object_updates $restore

### Set Ignore Sink Pins During CTS to Control Buffering ###
Source /project/moonwalk/implementation/physical/TCL/\ moonwalk_
ignore_pins.tcl

#To keep spares from getting moved
set_ccopt_property change_fences_to_guides false
```

```
set_ccopt_property max_fanout 50

### Controlling Useful Skew Delay Insertion by 5% ###


set_ccopt_property auto_limit_insertion_delay_factor 1.05
set_ccopt_property -target_skew 0.2

### Forces CCOPT to Use These Constraints ###
set_ccopt_property -constrains ccopt
create_ccopt_clock_tree_spec -immediate

ccoptDesign


timeDesign -expandedViews -numPaths 1000 -postCTS -outDir ../RPT/
cts -prefix CTS1


saveDesign ../MOONWALK/cts1.enc -compress


summaryReport -noHtml -outfile ../RPT/cts/cts0_summaryReport.rpt


### Clock Tree Timing Optimization with CTS Constraints   ###


### Reset All Ignore Sink Pins Used During CTS ###
source /project/moonwalk/implementation/physical/TCL\
/moonwalk_reset_ignore_pins.tcl

set_interactive_constraint_modes [all_constraint_modes -active]
set_propagated_clock [all_clocks]

set_interactive_constraint_modes [all_constraint_modes -active]
source  /project/moonwalk/implementation/physical/TCL/\
moonwalk_clk_gate_disable.tcl

set restore [get_global timing_defer_mmmc_object_updates]
set_global timing_defer_mmmc_object_updates true
set_analysis_view -update_timing
set_global timing_defer_mmmc_object_updates $restore

setOptMode -fixFanoutLoad true
```

```
setOptMode -addInstancePrefix CTS2_

optDesign -postCTS


timeDesign -postCTS -expandedViews -numPaths 1000 -outDir ../RPT/
cts -prefix CTS2


saveDesign ../MOONWALK/cts2.enc -compress


### Clock Tree Timing Optimization with Design Constraints  ###
set_interactive_constraint_modes [all_constraint_modes -active]

update_constraint_mode -name setup_func_mode \
   -sdc_files [list /project/moonwalk/implementation/physical/SDC/
moonwalk.sdc]
create_analysis_view -name setup_func \
   -constraint_mode setup_func_mode \
   -delay_corner slow_max

update_constraint_mode -name hold_func_mode \
   -sdc_files [list /project/moonwalk/implementation/physical/SDC/
moonwalk.sdc]
create_analysis_view -name hold_func\
                                      -constraint_mode hold_func_mode
\
                                        -delay_corner fast_min

set_analysis_view -setup [list setup_func] -hold [list hold_func]


set_interactive_constraint_modes [all_constraint_modes -active]
set_propagated_clock [all_clocks]

set restore [get_global timing_defer_mmmc_object_updates]
set_global timing_defer_mmmc_object_updates true
set_analysis_view -update_timing
set_global timing_defer_mmmc_object_updates $restore

setOptMode -fixFanoutLoad true
setOptMode -addInstancePrefix CTS3_
optDesign -postCTS
```

```
timeDesign -postCTS  -expandedViews -numPaths 1000 -outDir ../RPT/
cts -prefix CTS3
timeDesign -postCTS -hold -expandedViews -numPaths 1000  -outDir
../RPT/cts -prefix CTS3

saveDesign ../MOONWALK/cts3.enc -compress


### Fix Hold Violations Without Breaking Setup Timing  ##


set_interactive_constraint_modes [all_constraint_modes -active]
set restore [get_global timing_defer_mmmc_object_updates]
set_global timing_defer_mmmc_object_updates true
set_analysis_view -update_timing
set_global timing_defer_mmmc_object_updates $restore

setAnalysisMode -honorClockDomains true


setOptMode -addInstancePrefix HLD_FIX_


set_interactive_constraint_modes [all_constraint_modes -active]


source    /project/moonwalk/implementation/physical/TCL/moonwalk
clk_gate_disable.tcl


setOptMode \
  -fixHoldAllowSetupTnsDegrade false \
  -ignorePathGroupsForHold {in2reg reg2out in2out}

setDontUse dly1svtd1     false
setDontUse dly2svtd1     false
setDontUse dly3svtd1     false
setDontUse bufsvtdp5     false
setDontUse bufsvtd2p5  false

setOptMode -holdFixingCells {  bufd2p5 bufdp5 bufd2 dly1d1 dly2d1
}
```

```
optDesign -postCTS -hold


timeDesign -postCTS  -expandedViews -numPaths 1000 -outDir ../RPT/
cts -prefix CTS
timeDesign -hold -postCTS -expandedViews -numPaths 1000 -outDir
../RPT/cts -prefix CTS


saveDesign ../MOONWALK/cts.enc


summaryReport -noHtml -outfile ../RPT/cts/cts_summaryReport.rpt


if { [info exists env(FE_EXIT)] && $env(FE_EXIT) == 1 } {exit}
```

6.5   Summary

Chapter 6 discusses Clock Tree Synthesis (CTS) which is an important and challenging part of the physical design flow with respect to speed and power. It covers
the importance of on-chip variation (OCV) as well advanced OCV (ACOV) for
advanced node technologies.

The ideal result of CTS is uniform clock cell insertions while meeting design
timing requirements (setup and hold) for all clocks.

CTS, as presented in Chap. 6, is based on a divide-and-conquer strategy. Rather
than performing CTS all at once, CTS is broken into four different stages:

1. Building clock tree structures using modified design functional constraints to
   CTS constraints and guiding CTS by adding such constraints as ignore sink pin
   (i.e., flip-flop's clock port), false paths, disable timing arcs, etc. These constraints
   are used to stop CTS buffering at the clock ports (e.g., clock dividers). At this
   stage, the MMMC method is applied to fix both setup and hold timing violations
   simultaneously.
2. Timing optimization is performed during the second stage and is based on CTS
   constraints. As in stage 1, MMMC is applied to fix both setup and hold timing
   violations.
3. Stage 3 is the final CTS timing optimization based on the actual design constraints rather than CTS constraints. MMMC is used to fix all design setup and
   hold timing violations. Since MMMC flow gives priority to fixing setup timing
   violations over hold timing violations, at the end of this stage, there may be some
   hold timing violations which need to be fixed.
4. Stage 4 is to fix the remaining design hold timing violations explicitly without
   breaking the design setup timing.

Bibliography

1. K. Golshan, *Physical Design Essentials, an ASIC Design Implementation Perspective* (Springer Business Media, 2007) New York, USA
2. N.H. Weste, K. Eshraghian, *Principle of CMOS VLSI DESIGN, A Systems Perspective* (Addison-Wesley, Boston, 1985)
3. Cadence Design Systems, Inc. *Encounter Digital Implementation System* (2016)

Chapter 7
Final Route and Timing

> Success is stumbling from failure to failure with no loss of
> enthusiasm.
> Winston Churchill

After the completion of floorplan, placement, and CTS, the next phase is to perform the final routing and design timing closure across all functional modes and sub-modes (e.g., testing and input/output with multiple process corners). This is done in order to minimize the ECO effort required during the signoff phase.

During the final route and timing phase of physical design implementation, it is critical to meet all design timing requirements as well as its manufacturability. This is especially true for advanced node technologies as advanced node technologies have more design rules in comparison to larger technologies (i.e., above 40 nm).

ASIC designs are becoming increasingly more complex as the increased number of cells and their interconnections require a routing area larger than the standard cell area. This makes routing more difficult. If the prerequisite of routing is not fulfilled properly, routing can fail to complete or, at minimum, take an unacceptable amount of execution run time.

The prerequisites which influence the route-ability while meeting design timing requirements are standard cell design, well-prepared floorplan, quality of standard cell placement, and CTS as discussed in the previous chapters.

As mentioned in Chap. 2, in a typical physical implementation flow, one would concentrate on only the main functional mode of the design during physical design and relied on the ECO to fix design timing violations for other modes.

The objective presented here is to not only route the design without any physical violations (i.e., shorted nets) but to also close timing for all design modes utilizing the MMMC method in order to minimize design ECO efforts.

For larger technology nodes, routing is much easier because of large metal widths and their spacing as well as larger standard cell sizes. However, with advanced nodes (i.e., below 40 nm), routing has become more challenging due to the small metal widths and their spacing. In addition, standard cell sizes are much smaller.

© Springer Nature Switzerland AG 2020   139
K. Golshan, *The Art of Timing Closure*,
https://doi.org/10.1007/978-3-030-49636-4_7

Because there are more standard cells and interconnects, one could image the cumulative wire length for technologies using advanced nodes could easily exceed over 1 kilometer.

In addition, electromigration (EMI), which is caused by excessive current density, crosstalk coupling for adjacent nets (i.e., smaller metal net spacing along with their high rate of clock switching), leakage optimization, and the phenomena known as design for manufacturability (DFM) must be considered.

7.1 Electromigration Concerns

Most ASIC chips must have an MTTF (mean time to fail) of at least 10 years. However, this rate can be much lower for advanced nodes because of their smaller metal width due to higher metal resistance. The lower MTTF is due to current density and EMI of a given wire and is expressed by Black's equation:

$$\text{MTTF} = \frac{A}{J^2} \exp\left(\frac{E_a}{kT}\right) \qquad (7.1)$$

where A is metal constant, J is current density (i.e., the number of electrons crossing a unit area per unit time), k is the Boltzmann constant, E_a is the activation energy, and T is the temperature.

Based on Eq. (7.1), the MTTF due to EMI depends on two parameters—temperature and current density. One should note that the MTTF is not an instant failure. Rather, it requires some time to occur.

High current density causes the electrons in the metal to move at an accelerated speed. This is known as electron wind. These electrons transfer their momentum to other atoms, and the atoms get displaced from their original position and exceed EMI current density limits.

With advanced node technology, the probability of failure due to EMI increases because both the power density and the current density are increasing. Specifically, metal layer line widths and wire cross-sectional areas will continue to decrease over time.

Currents are also reduced due to lower supply voltages and shrinking gate capacitances. However, as voltage and current reduction are constrained by increasing frequencies, the more marked decrease in cross-sectional areas will give rise to increased current densities in advanced node ASICs. Over time, increases of current density will cause EMI which will then lead to creating voids and hillocks. Hillocks will create shorts between the lower and upper metal. Voids are an opening of hole in the metal layers.

In general, the root causes for EMI are power routing, nonuniform net connections, high-frequency switching nets, and IR drop (i.e., VDD drop plus VSS rise).

Figure 7.1 shows the high current density occurring when the electrical current moves from wide metal to a thinner area with respect to power supply connections (e.g., VDD).

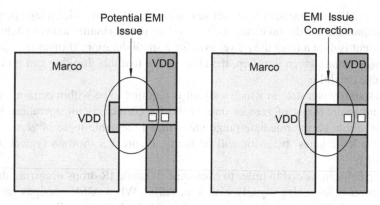

Fig. 7.1 Illustration of electromigration issue with power (VDD) net

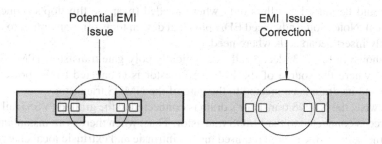

Fig. 7.2 Illustration of electromigration issue with segmented net

Figure 7.2 shows a potential issue with a segmented net such as a wide metal segment connection to a narrower metal layer segment (e.g., clock net with a non-default rule and a regular rule of metal layer).

The solution for eliminating EMI issues includes:

- Increasing the width of the wire (power, ground, and clock nets)
- Insertion of buffers (clock and regular net)
- Upsizing the driver cell (clock and regular net)
- Switching the net into a higher metal layer (lower resistance Cu vs. Al)
- Reducing IR drop by improving IC power and ground routing

7.2 De-coupling Capacitance Cell Consideration

De-coupling capacitance cells (i.e., decap cells) are temporary capacitors added in the design between power and ground rails in order to counter functional failures due to dynamic IR drop.

Dynamic IR drop happens at the active edge of the clock at which a high percentage of sequential and digital elements switch. Due to this simultaneous switching, a high current is drawn from the power grid for a small duration. If the power source is far away from a given flop-flop, the chances are that this flop-flop can go into a metastable condition.

Metastability is a state in which a signal is required to be within certain voltage or current limits (logic of zero or one levels) for correct circuit operation. If the signal is within an intermediate range other than acceptable logic of zero or one levels, the logic gates' behavior will be faulty. Figure 7.3 shows a typical decap cell design.

Decap cells are added in order to overcome dynamic IR drops occurring during simultaneously switching signals within an ASIC. When AISIC designs have an active edge of clock and the current requirement is high, decap cells discharge and provide boost to the power grid. Since decap cells have additional leakage current, they should be placed as fillers only where needed (near the flip-flop's sequential instances). Note: Some advanced EDA physical design tools offer an option to automatically insert decap cells where needed.

As shown in Fig. 7.3, decap cells are typically poly-gate transistors (PMOS and NMOS) where the source of the PMOS transistor is connected to the power rail (VDD) and its drain is connected to the gate of the NMOS transistor.

Likewise, the NMOS transistor's drain is connected to the ground (VSS) rail, and its source is connected to the PMOS transistor. Thus, when there is an instantaneous switching activity, the charge required moves intrinsic and extrinsic local charges to reservoirs as opposed to moving them to voltage sources.

Extrinsic capacitances are decap cells placed in the design. Intrinsic capacitances are those present naturally in the circuit, such as the grid capacitance, the variable capacitance inside nearby logic, and the neighborhood loading capacitance exposed when the PMOS or NMOS transistor channels are open.

As mentioned before, one drawback of decap cells is that they are very leaky, so the more decap cells used, the more leakage. Another drawback, which many designers ignore, is the interaction of the decap cells with the package RLC network.

Since the die is essentially a capacitor with very small R (resistance) and L (inductance) and the package is a huge RLC network, the more decap cells placed, the more chance of turning the circuit into its resonance frequency. This is problematic since both VDD and VSS will then be oscillating.

Fig. 7.3 De-coupling
capacitance cell design

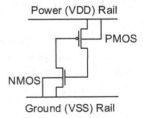

It is not uncommon that some physical designers tend to place decap cells near high activity clock buffers. However, it is recommended to use decap optimization flow (i.e., dynamic power analysis) in order to understand charge requirements at every moment in time and to figure out how much decap to place at any node. This should be done while taking package models (i.e., package RLC network) into account to ensure resonance frequency is not affected.

7.3 Engineering Change Order Metal Cells

In today's ASIC designs with multimillion gates, along with the increase in design complexity, the chance of functional ECOs is equally increased.

Metal ECOs play a vital role in absorbing last-minute design changes before taping out and hence help save millions in terms of mask cost. They also help avoid all-layer mask changes in respins to fix bugs found in the design validation. Though design fixes belong to the logical domain, physical aspects of implementation are poorly understood. This section will focus on metal ECO implementation methodologies with an emphasis on mask programmable cells.

Originally the idea of mask programmable cells was gate-array technology. The gate-array technology is an approach to the manufacturing of ASICs using a prefabricated chip with components that are later connected into logic devices (e.g., NAND gates, etc.) by adding metal interconnect layers at IC foundries.

Spare cells are inserted in the physical design in the form of spare modules, as discussed in Chap. 4. A spare module mostly includes a wide variety of combinational and sequential cells (e.g., inverter, buffer, multiplexer, flip-flop, etc.) from the standard cell library. The input of these cells is tied to either power or ground in order to insure no floating outputs.

This approach has some inherent probabilistic problems. In order to resolve, the metal programmable ECO cells also need to be inserted during the final stage of routing. The additional programmable ECO cells are beneficial during the design ECO process. High drive strength programmable ECO cells are needed in order to drive long nets and avoid timing violations that are due to large capacitance loading.

There are two types of metal programmable ECO cells. One is ECO filler (as shown in Fig. 7.4) and the other is functional ECO cells.

The ECO filler cells are constructed based upon the base layers known as frontend-of-line (FEOL). FEOL are implant, diffusion, and poly-layers. This allows any functional ECO to be performed using back-end-of-line layers.

Functional programmable ECO cells include a wide variety of combinational and sequential cells with multiple drive strengths realized by using width multiples for filler cells. Their cell layout has the same FEOL footprint as that of ECO filler cells. The only difference is that the functional ECO cell will use ECO filler FEOL layout and have contact connections to poly-layers and diffusion and *metal1* layers for internal connections in order to construct a functional gate.

Fig. 7.4 Example of metal
ECO filler spare cell

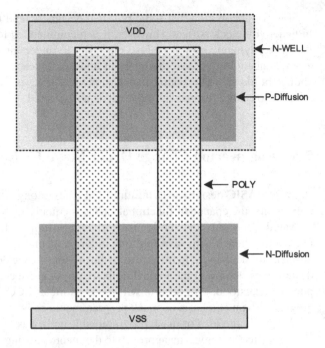

Fig. 7.5 Example of metal ECO functional spare cell

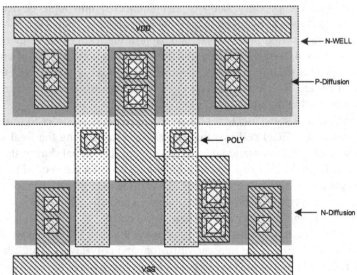

Figure 7.5 shows a functional ECO cell with the same footprint as the ECO filler cell shown in Fig. 7.4.

Like floorplan, placement, and CTS flows, the routing flow will take several steps to complete. At the complication of final route, the objective is to have a physically clean route which meets all design rules, as well as timing closure, for all functional and test modes in the design.

The steps for final routing are:

- Initial design routing
- Design routing optimization
- Leakage and optional (SI and DFM) optimizations
- Invoking MMMC and reporting timing for all modes
- Applying manual ECO to remove any large timing violations
- Applying MMMC to fix remaining timing violations
- Final routing

7.4   Initial Design Routing

The objective of initial design routing is to route and time the design, fixing setup timing violations. One should note that today's routers will not have any open nets, and therefore, open nets will not be discussed.

After completion of the initial design routing, there should not be any shorted nets. If there are any shorted nets, most EDA tools have a "search and repair" option to resolve. This option only works well on an isolated area with shorted net violations. In addition, this option may move the violations from one area to another. This process of "search and repair" may become problematic when there are multiple shorted net violations.

If there remains shorted net violations after using the "search and repair" option, use the procedure (proc) to remove. It should be noted that it may require several attempts to remove shorted net violations. This approach automatically identifies the shorted nets and creates a list. In addition, it automatically deletes these shorted nets from the design.

Once the shorted nets are deleted, ECO and design routing need to be performed. In doing so, there is a procedure called "reroute_shorts." This procedure can be found under *moonwalk_config.tcl* in Chap. 4.

It is not uncommon to have many shorted nets remaining even after a few attempts of removing the violations. This could be a result of routing congestions. Often the congested area in the design results from clock net routing using non-default rules (NDR) during CTS.

The procedure `remove_ndr_nets` removes clock routing NDRs for a specified clock net, thus relaxing the routing area. However, caution should be exercised in using this command as it may introduce problems such as capacitance coupling in the clock routing.

Another problem in routing congestions may be caused by a lack of enough area for routing (i.e., high routing utilization). This is especially true for advanced nodes where the standard cell area utilization is high due to their small standard cell's geometry, thus not enough area for routing.

Having 12 routing layers is not uncommon for advanced nodes. To decide whether chip area needs to be increased or routing layers added, one needs to do a manufacturing cost analysis to see which option is least expensive.

Scenic net routing, which is a by-product of high routing utilization, impacts timing. This problem can be fixed by either increasing the chip area or adding more routing layers.

The following are steps, and options are required for the initial and final route:

```
set_analysis_view -setup [list setup_func] -hold [list hold_func]
```

```
setNanoRouteMode -routeTopRoutingLayer 12
```

LithoDriven refers to lithography awareness routing and mainly applies to the advanced node process:

```
setNanoRouteMode -routeWithLithoDriven true
```

Since at this stage the objective is to route the design without any shorts, it is recommended to set TimingDriven to false:

```
setNanoRouteMode -routeWithTimingDriven false
```

Signal integrity (SI) is the ability of an electrical signal to reliably carry information and resist the effects of high-frequency electromagnetic interference from nearby signals. SiDriven option is set to reduce crosstalk and its delta delay.

Crosstalk is due to the influence of cross coupling capacitance, which is from switching the signal from one net (aggressor) to the neighboring net (victim).

Setting the route with SiDriven option true allows the physical design tool to increase the spacing between the aggressor net and victim net, so that the cross-coupling capacitance decreases as spacing increases. This then reduces the effect of crosstalk. In addition, the physical design tool inserts buffers in order to boost the strength of the victim net and, thereby, reduce the effect of crosstalk.

Another technique to avoid crosstalk is to place a shielding which is the ground net (e.g., VSS) between the aggressor and victim net so that the voltage will discharge through the ground net. However, this also increases the sidewalk capacitance, which may in turn impact the net delay timing.

Glitch or noise bump occurs when a transition on an adjacent signal (aggressor net) causes a noise bump/glitch on the constant signal (victim net). A noise model is required in order to fix glitches or noise bumps. Most of the time, there are separate dedicated EDA tools for noise analysis.

The following shows an example of the SI option:

```
setNanoRouteMode -routeWithSiDriven true
setSIMode -deltaDelayThreshold 0.01 \
                -analyzeNoiseThreshold 80 -fixGlitch false
```

Below is an example of how to set the analysis mode with OCV and CPPR (as discussed in Chap. 6) with setup timing only:

```
setAnalysisMode -analysisType onChipVariation -cppr setup
```

After these options are set, the initial design routing is completed by routeDesign.

After initial routing, there should not be any shorted net violations or large setup timing violations. One must time the design and generate and review the timing violation report before continuing to the next stage.

7.5   Design Routing Optimization

Once the initial routing is confirmed (i.e., no shorted nets and no large setup timing violations), the next step is routing optimization.

At this stage, the objective is to address both setup and hold violations, if any. To do this, the analysis mode needs to be set for both setup and hold with OCV and CPPR options:

```
setAnalysisMode -analysisType onChipVariation -cppr both
```

It should be noted that analysis mode is set for setup and hold (both) timing. In addition, one needs to make sure all clocks are set in propagation mode.

During routing optimization, DelayCalMode is set for SIAware, and ExtractRCMode is activated with capacitance coupling effects:

```
setDelayCalMode -SIAware true -engine default
setExtractRCMode -engine postRoute -coupled true\
                                -effortLevel medium
```

To complete, one needs to optimize the routing. Since the design is already routed, the postRoute option is used during design routing optimization:

```
optDesign -postRoute -setup
```

Another routing optimization step is to fix any remaining hold timing violations. If the standard cell's design library supports low leakage cells (e.g., with LL prefix) and they were set with "do not use" attributes because of their slow intrinsic delay, it is recommended to use them for the hold timing violation fix. Delay cells in the standard cell library should also be used to fix hold timing violations:

```
foreach cell [list  {LL */*} ] { setDontUse $cell false }
```

```
setDontUse dly2d1 false
setDontUse dly3d1 false
```

Moreover, it is recommended to use a selection of cells from the design's standard cell library for fixing hold timing violations:

```
setOptMode -holdFixingCells { LLbufd2 LLbufd3 dly2d1 dly3d1 }
```

Fixing the remaining hold timing violations is performed with post route and hold option:

```
optDesign -postRoute -hold
```

Often physical design starts with using HVT cells for the purposes of addressing overall ASIC design leakages. In addition, other cell types, such as SVT and LVT, are attributed with "do not use" attributes because of their high leakage. However, from a performance timing point of view, HVT cells are slower than SVT and LVT cells. Therefore, it may be necessary to use SVT and LVT cells in addition to HVT cells to fix the remaining setup timing violations.

To allow EDA tools to use these cells, the "do not use" attribute needs to be removed during the fixing of setup timing violations:

```
foreach cell [list {svt*/* lvt*/*} ] { setDontUse $cell false }
```

After removing "do not use" attributes, route optimization is performed by the following command:

```
optDesign -postRoute -setup
```

The last stage of routing optimization is to minimize design leakage power. The EDA tools will replace cells with high leakage (e.g., LVT and SVT) with low leakage cells (HVT) as long as swapping these cells does not break design setup and hold timing. One can report timing before and after leakage optimization. The leakage optimization is performed by optLeakagePower:

```
report_power -leakage
optLeakagePower
report_power -leakage -power_domain
```

There are additional optimizations which can be performed during routing optimizations such as SI and DFM.

DFM consists of a set of different physical design rules, known as "recommended design rules," regarding the shapes and polygons (e.g., standard cell layout). DFM rules apply to spacing/width of interconnect layers, via/contact overlaps, and contact/via redundancy.

DFM rules are applied to minimize the impact of physical process variations on performance and other types of parametric yield loss. For example, all different types of worst-case simulations are essentially based on a set of worst-case device parameters that are intended to represent the variability of transistor performance over the full range of variation in the fabrication process.

Changing the spacing and width of the interconnect wires requires a detailed understanding of yield loss mechanisms, as these changes trade off against one another. For example, introducing via redundancy will reduce the chance of via problems during the manufacturing process and reduces via resistance. Whether this is a good idea or not is dependent upon the details of the yield loss models and the characteristics of the ASIC design.

For advanced nodes (e.g., below 20 nm) where the manufacturing processes are undergoing process improvements, it is essential to add DFM rules as much as possible as long as it does not impact the size of the chip area.

The details of SI and DFM optimizations are outlined in the *moonkwalk_frt.tcl* final route in the **Final Routing Script** section of this chapter.

7.6 MMMC Design Timing Closure

The initial routing and optimizations discussed thus far apply only to the main functional mode. In order to close all remaining setup and hold timing violations on other modes in the design, MMMC methodology will be utilized.

For the purpose of illustration, functional scan capture and scan shift modes are set for the analysis mode:

```
set_analysis_view -setup [list setup_func] \
                        -hold  [list hold_func hold_scanc
hold_scans]
```

Once analysis modes are set, they need to be defined. In this example, the modes are setup_func_mode, hold_func_mode, hold_scanc_mode and hold_scans_mode:

```
update_constraint_mode -name setup_func_mode \
  -sdc_files [list \ /project/moonwalk/implementation/physical/SDC/
moonwalk_func.sdc]
create_analysis_view -name setup_func \
   -constraint_mode setup_func_mode \
   -delay_corner slow_max
```

It should be noted that for scan only, hold timing analysis is applicable and setup timing analysis is not:

```
update_constraint_mode -name hold_func_mode \
  -sdc_files [list \ /project/moonwalk/implementation/physical/SDC/
moonwalk_func.sdc]
create_analysis_view -name hold_func\
   -constraint_mode hold_func_mode \
   -delay_corner fast_min

update_constraint_mode -name hold_scanc_mode \
  -sdc_files [list \ /project/moonwalk/implementation/physical/SDC/
moonwalk_scanc.sdc]
create_analysis_view -name hold_func\
   -constraint_mode hold_func_mode \
   -delay_corner fast_min

update_constraint_mode -name hold_scans_mode \
  -sdc_files [list \ /project/moonwalk/implementation/physical/SDC/
moonwalk_scans.sdc]
create_analysis_view -name hold_func\
   -constraint_mode hold_func_mode \
   -delay_corner fast_min
```

These modes are defined in the MMMC view definition file (i.e., *monnwalk_view_def.tcl*).

It is important to note that in today's ASIC design, there are more than one functional (e.g., UART, USB) and test (e.g., MBIST) mode that can be included in setup and hold timing analysis view lists as shown in the below example:

```
set_analysis_view -setup [list setup_func setup_uart setup_mbist]
\
          -hold  [list hold_func hold_uart hold_mbist hold_scanc
hold_scans]
```

In order to understand timing violations across all modes in the design, it is recommended to time the design and generate and review timing reports for both setup and hold timing with post route options:

```
timeDesign -postRoute -setup  -numPaths 1000 \
                                        -outDir ../RPT/
mmmc -expandedViews


timeDesign -postRoute -hold  -numPaths 1000 \
                                        -outDir ../RPT/
mmmc -expandedViews
```

Before applying post route optimization, check both setup and hold timing violations for all modes in the design. If there are any large violations (e.g., greater than 300 ps), they will need to be reduced through manual ECOs.

For example, it is not uncommon to have large hold timing violations for scan capture and shift. Fixing scan shift hold timing violations is very straightforward from the ECO point of view. Delay or buffer cells need to be added to the scan input port of the violated flip-flops (e.g., scan input of flip-flop or SI) in order to fix scan shift violations without affecting any functional setup and hold timing. The reason for this is that all flip-flops in the design form back-to-back (i.e., scan chain) bypassing all combinational logics between them. Thus, by construction, there will be no setup timing violations among these flip-flops in the design.

Figure 7.6 shows an example of fixing hold timing violations during scan shift. It should be noted that two buffers were added to SI input of INST2 in order to resolve hold timing violations due to the scan clock skew between INST1 and INST2.

However, fixing scan capture hold timing violations is not as straightforward as scan shift ECOs.

During scan capture, the combinational logics between two given flip-flops are not bypassed. This is so manufacturing faults in the design will be captured.

A problem arises during the scan capture hold timing fix when scan clock is used as a functional clock. Setup and hold timing are fixed in functional mode using multiple clocks. Since a single clock is used during the scan capture hold timing fix, functional setup timing may break. Then when trying to fix functional setup timing, the scan capture hold timing may break.

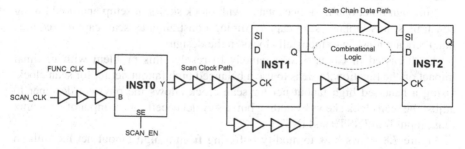

Fig. 7.6 Example of hold timing violation fix during scan shift

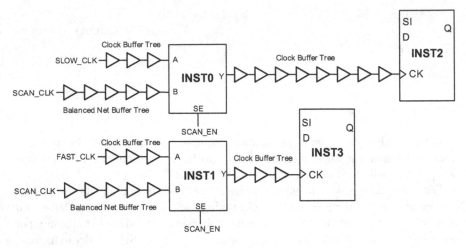

Fig. 7.7 Example of scan and functional hold timing conflicts

During scan capture, there is only one single clock (i.e., scan clock) which is used to exercise all combinational logics to determine manufacturing faults. This leads to clock skew among different functional clocking which leads to scan capture hold timing violations.

Figure 7.7 shows the conceptual issue with functional and scan clock during scan capture mode.

Adding delay or buffer cells to the data inputs of violated flip-flops (e.g., D input) is often used to fix scan capture hold timing violations. However, this may increase the possibility of breaking setup timing in the functional clocking modes (e.g., slow clock and fast clock).

In the typical physical design approach, this may require unnecessary manual ECOs in order to fix both hold (scan capture) and setup (functional) timing violations due to differences in skews impacting functional clock tree insertions.

In the example shown in Fig. 7.7, during functional mode (SLOW_CLK), there are 11 clock tree buffers, and for the other clock (FAST_CLOCK), there are only 6 clock tree buffers.

Although functionality is acceptable with clock skews in setup and hold timing constraints, this violates scan capture timing constraints as scan capture requires that there are no skews among all clocks in the design.

As discussed in Chap. 5, one remedy to resolve this problem with minimal impact on the functional side is to use a balanced high fanout net fix for scan clock. Using a balanced high fanout net for scan clock allows the physical designer to adjust the scan clock skews by adding buffers or delay cells at the input of scan mux (e.g., input B of INST0 and INST1).

Figure 7.8 shows how to modify buffering from a high fanout net fix with an option of a balanced net during the placement stage for scan clock muxs (i.e., their B inputs), in comparison to Fig. 7.7, without impacting functional clock tree structures (i.e., adding buffers to D inputs of flip-flops after Y outputs of scan muxs).

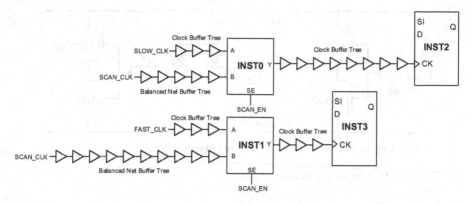

Fig. 7.8 Example of manual ECO for scan clock balancing

Comparing Fig. 7.7 (which is before ECO) and Fig. 7.8 (which is after ECO), one can see that five buffers were added to the input B of scan mux (INST1) in order to minimize scan clock skew between flip-flops with slow and fast functional clocks. This was done without impacting functional clock tree structures. By counting the number of scan clock paths, one can see there are no scan capture violations for either flip-flop (i.e., INST2 and INST3) as they have identical delays and minimal skews for the scan clock (SCAN_CLK).

Once all large setup and holds are removed, perform the post route optimization with post route options like those discussed in the **Design Routing Optimization** section. The only difference between the two is that MMMC is activated at this step of design timing optimization.

The final steps are to export full interconnect RC extractions for each process corner in the design (e.g., slow and fast process), post route (often referred to as CTS netlist) for the purpose of STA, gate-level simulations, IR drop, dynamic power, and noise analysis.

The final step after the filler cells (i.e., decap cells and metal ECO cells) are inserted is to verify design connectivity, geometry, and antenna to see if there any issues. All issues need to be resolved.

The most common technique used to resolve antenna problems is to reduce the peripheral metal length that is attached to the transistor gates. This is accomplished by segmenting the wire of one metal type to several segments of a different metal type and connecting these various types of metals through via connections (shown in Fig. 7.9). Another way to resolve would be to insert protection diodes.

It is important to realize that these extra via insertions, done while repairing antenna violations, increases the wire resistance as the via is highly resistive. Therefore, it is strongly recommended to extract the parasitic routing after resolving all antenna violations in order to account for the extra resistance.

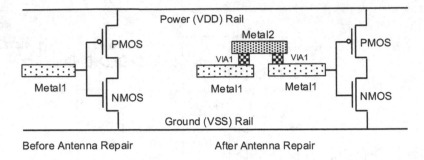

Before Antenna Repair   After Antenna Repair

Fig. 7.9 Antenna metal repair procedure

7.7 Final Routing Script

Final routing and timing optimization script:

```
### Setup Environment ###
source    /project/moonwalk/implementation/physical/TCL/moonwalk_
config.tcl

### Initial Design Routing     ###
restoreDesign ../MOONWALK/cts.enc.dat moonwalk
source    /project/moonwalk/implementation/physical/TCL/moonwalk_
setting.tcl

generateVias


### Turn Off SVT/LVT Cells for Routing ( i.e. high leakage) ###
### These Options Could be Turned ON if Performance is Required ###

foreach cell [list {svt*/*} ] { setDontUse $cell true }
foreach cell [list {lvt*/*} ] { setDontUse $cell true }

### Turn Off Low Leakage Cells for Routing Due High Intrinsic Delay
###
foreach cell [list  {LL_*/*} ] { setDontUse $cell true }

set_interactive_constraint_modes [all_constraint_modes -active]


set_analysis_view -setup [list setup_func] -hold [list hold_func]
```

```
setNanoRouteMode -routeTopRoutingLayer 12
setNanoRouteMode -routeWithLithoDriven false
setNanoRouteMode -routeWithTimingDriven false
setNanoRouteMode -routeWithSiDriven true
setNanoRouteMode -droutePostRouteSpreadWire false

setSIMode    -deltaDelayThreshold   0.01   -analyzeNoiseThreshold
80 -fixGlitch false


setOptMode -addInstancePrefix INT_FRT_


set active_corners [all_delay_corners]
setAnalysisMode -analysisType onChipVariation -cppr setup

set_interactive_constraint_modes [all_constraint_modes -active]
source     /project/moonwalk/implementation/physical/TCL/moonwalk_
clk_gate_disable.tcl

routeDesign
timeDesign  -postRoute  -prefix  INT_FRT  -expandedViews  -numPaths
1000 -outDir ../RPT/frt

saveDesign ../MOONWALK/int_frt.enc -compress


### Design for Manufacturing DFM (optional) ###
#setNanoRouteMode  -droutePostRouteSpreadWire  true  -routeWith-
TimingDriven false
#routeDesign -wireOpt
#setNanoRouteMode -droutePostRouteSwapVia multiCut
#setNanoRouteMode -drouteMinSlackForWireOptimization <slack>
#routeDesign -viaOpt
#setNanoRouteMode  -droutePostRouteSpreadWire  false  -routeWith-
TimingDriven true

### Design Routing Optimization ###
setNanoRouteMode -drouteUseMultiCutViaEffort medium
setAnalysisMode -analysisType onChipVariation -cppr both
setDelayCalMode -SIAware true -engine default
setExtractRCMode  -engine  postRoute  -coupled  true  -effortLevel
medium
```

```
set_interactive_constraint_modes [all_constraint_modes -active]
set_propagated_clock  [all_clocks]

setOptMode -addInstancePrefix FRT_


optDesign -postRoute -setup  -prefix FRT


setOptMode -addInstancePrefix FRT_HOLD_


### Turn on Low Leakage Cells  for Hold Timing Optimization ###
foreach cell [list  {LL_*/*} ] setDontUse $cell false }

source    /project/moonwalk/implementation/physical/TCL/moonwalk_
setting.tcl


setDontUse dly2d1 false
setDontUse dly3d1 false

setOptMode -holdFixingCells  { LLbufd2  LL_bufd3  LL_bufd4  dly2d1
dly3d1 }


set_interactive_constraint_modes [all_constraint_modes -active]
source    /project/moonwalk/implementation/physical/TCL/moonwalk_
clk_gate_disable.tcl

optDesign -postRoute -hold


timeDesign -postRoute -hold -prefix OPT_FRT -expandedViews -numP-
aths 1000 \
 -outDir ../RPT/frt
timeDesign  -postRoute  -prefix  OPT_FRT  -expandedViews  -numPaths
1000\
 -outDir ../RPT/frt


saveDesign -tcon ../MOONWALK/opt_frt.enc -compress


### Turn on SVT/LVT Cells for Timing Optimization its was Turned
Off at Initial  ###
```

```
foreach cell [list  {svt*/*} ] { setDontUse $cell false }

source     /project/moonwalk/implementation/physical/TCL/moonwalk_
setting.tcl
set_interactive_constraint_modes [all_constraint_modes -active]
source     /project/moonwalk/implementation/physical/TCL/moonwalk_
clk_gate_disable.tcl

set_interactive_constraint_modes [all_constraint_modes -active]
set report_timing_format {instance cell pin arc fanout load delay
arrival}
set_propagated_clock [all_clocks]

optDesign -postRoute -setup


### Leakage Optimization  ####
report_power -leakage
optLeakagePower
report_power -leakage -power_domain all -outfile ../RPT/power.rpt

### SI Optimization (Optional)  ####
#set_interactive_constraint_modes [all_constraint_modes -active]
#setAnalysisMode -analysisType onChipVariation -cppr both
#setDelayCalMode -SIAware false -engine signalstorm
#setSIMode -fixDRC true -fixDelay true -fixHoldIncludeXtalkSetup
true -fixGlitch false
#setOptMode -fixHoldAllowSetupTnsDegrade false -ignorePathGroups-
ForHold {reg2out in2out}
#setOptMode -addInstancePrefix FRT_SI
#optDesign -postRoute -si
#timeDesign -postRoute -si -expandedViews -numPaths 1000 -outDir
../RPT/frt -prefix FRT_SI
#setAnalysisMode -honorClockDomains true
#setOptMode -addInstancePrefix FRT_SI_HOLD

#optDesign -postRoute -si -hold


#timeDesign -postRoute -si -hold  -prefix FRT_SI -expandedViews -num-
Paths 1000\
                                              -outDir ../RPT/frt
#timeDesign -postRoute -si  -setup -expandedViews -numPaths 1000\
                                              -outDir ../RPT/frt -prefix
FRT_SI
```

```
#saveDesign ../MOONWALK/frt_si.enc -compress

### Completing the Routing Stage ###
setNanoRouteMode -droutePostRouteLithoRepair false

setNanoRouteMode -drouteSearchAndRepair true
globalDetailRoute

timeDesign -postRoute -hold -prefix FRT -expandedViews -numPaths
1000 -outDir ../RPT/frt
timeDesign -postRoute  -setup -prefix FRT -expandedViews -numPaths
1000 -outDir ../RPT/frt

saveDesign -tcon ../MOONWALK/frt.enc -compress


### Delete Empty Modules During Netlist Optimization to prevent
Physical Verification issue ###
deleteEmptyModule
saveDesign -tcon ../MOONWALK/frt.enc -compress

setFillerMode -corePrefix FILL -core "Add List of Filler cells  :
metal ECO, Decap and Filler Cells"
addFiller

verifyConnectivity -noAntenna
verifyGeometry
verifyProcessAntenna

saveDesign -tcon ../MOONWALK/moonwalk.enc -compress


summaryReport -noHtml -outfile ../RPT/frt/moonwalk_summaryReport.
rpt


### Generating an Area Report by Hierarchical Module ###
#reportGateCount -limit 1  -level 8 -outfile ../RPT/frt/moonwalk_
areaReport.rpt

defOut  -floorplan  /project/moonwalk/implementation/physical/def/
moonwalk_flp.def
if { [info exists env(FE_EXIT)] && $env(FE_EXIT) == 1 } {exit}
```

The multi-mode multi-corner (MMMC) script:

```
#### Setup Environment ###
source    /project/moonwalk/implementation/physical/TCL/moonwalk_
config.tcl

#### Multi Mode Multi Corner (MMMC) Setup ###
restoreDesign ../MOONWALK/frt.enc.dat moonwalk
source /project/moonwalk/implementatio/physical/TCL/moonwalk_set-
ting.tcl

generateVias


saveDesign ../MOONWALK/frt_no_mmc_opt.enc  -compress


### Uncomment Before Applying MMMC ECO ###
#setEcoMode -honorDontUse true -honorDontTouch true -honorFixed-
Status true
#setEcoMode -refinePlace true -updateTiming true -batchMode false

set_analysis_view -setup [list setup_func] -hold [list hold_func
hold_scanc hold_scans]


set_interactive_constraint_modes [all_constraint_modes -active]
update_constraint_mode -name setup_func_mode \
    -sdc_files [list /project/moonwalk/implementation/physical/SDC/
moonwalk_func.sdc]
create_analysis_view -name setup_func \
    -constraint_mode setup_func_mode \
    -delay_corner slow_max

update_constraint_mode -name hold_func_mode \
    -sdc_files [list /project/moonwalk/implementation/physical/SDC/
moonwalk_func.sdc]
create_analysis_view -name hold_func\
    -constraint_mode hold_func_mode \
    -delay_corner fast_min

update_constraint_mode -name hold_scans_mode \
    -sdc_files [list /project/moonwalk/implementation/physical/SDC/
moonwalk_scans.sdc]
create_analysis_view -name hold_scans \
    -constraint_mode hold_scans_mode \
    -delay_corner fast_min
```

```
update_constraint_mode -name hold_scanc_mode \
   -sdc_files [list /project/moonwalk/implementation/physical/SDC/
moonwalk_scanc.sdc]
create_analysis_view -name hold_scanc \
   -constraint_mode hold_scanc_mode \
   -delay_corner fast_min

set_analysis_view -setup [list setup_func ] \
                  -hold [list hold_func hold_scanc hold_scans ]

set_interactive_constraint_modes [all_constraint_modes -active]
source /project/moonwalk/implementatio/physical/TCL/moonwalk_clk_
gate_disable.tcl

set_interactive_constraint_modes [all_constraint_modes -active]
set report_timing_format {instance cell pin arc fanout load delay
arrival}
set_propagated_clock [all_clocks]

### Generate Setup and Hold Timing Report For Analysis Before
Applying MMMC ###
timeDesign -setup -postRoute -numPaths 1000 -outDir ../RPT/
mmc -expandedViews
timeDesign -hold   -postRoute -numPaths 1000 -outDir ../RPT/
mmc -expandedViews

setOptMode -addInstancePrefix MMC_


optDesign -postRoute
timeDesign -postRoute -setup -numPaths 1000 -outDir ../RPT/
mmc -expandedViews
timeDesign -hold -postRoute -numPaths 1000 -outDir ../RPT/
mmc -expandedViews

setDontUse dly2d1 false
setDontUse dly3d1 false

setOptMode -holdFixingCells { LL_bufd2 LL_bufd3 LL_bufd4 dly2d1
dly3d1


setOptMode -fixHoldAllowSetupTnsDegrade false -ignorePathGroups-
ForHold {reg2out in2out}
```

```
optDesign -hold -postRoute
optDesign  -setup -postRoute
timeDesign   -postRoute   -hold   -outDir  ../RPT/mmc   -numPaths
1000 -expandedViews
timeDesign   -postRoute   -setup   -outDir  ../RPT/mmc   -numPaths
1000 -expandedViews

saveDesign -tcon ../MOONWALK/frt.enc -compress
saveDesign -tcon ../MOONWALK/moonwalk.enc -compress
```

7.8 Summary

Chapter 7 discusses the final route and timing optimizations. In addition, it describes how to take advantage of multi-mode multi-corner (MMMC) methodology for all functional and testing modes simultaneously in order to minimize ECO activities. The detail of the final route process is also discussed—initial design routing, design routing timing optimization, and MMMC timing closure.

Additional topics addressed are electromigration concerns and how to avoid, decoupling capacitance cell design and their usage, engineering change order (ECO) using metal programmable cells in order to avoid changing front-end-of-line (FEOL) layers, and general cell design using back-end-of-line layers for changing logical functions.

Examples of complete final routing and MMMC scripts are included.

Bibliography

1. K. Golshan, *Physical Design Essentials, an ASIC Design Implementation Perspective* (Springer Business Media, 2007) New York, USA
2. J. Lienig, M. Thiele, The pressing need for electromigration-aware physical design, in *Proceedings of the International Symposium on Physical Design (ISPD)* (March 2018)
3. J.R. Black, Electromigration—a brief survey and some recent results, in *Proceeding IEEE International Reliability Physics Symposium* (December 1968)

Chapter 8
Design Signoff

> *The ultimate authority must always rest with the individual's*
> *own reason and critical analysis.*
> Dalai Lama

An ASIC design's signoff is the last phase of implementation. It requires verification and validation before committing to the silicon manufacturing process, which is commonly known as design tape-out.

Verification and validation are of paramount importance during signoff. Any failure during these phases not only impacts time to market but will also be very costly. This is especially true for advanced node technology where silicon processing is very expensive and time-consuming.

ASIC design verification is a process, which happens before tape-out, in which a design is tested (or verified) against its design specifications. This happens along with the development of the design and can start from the time the design architecture/micro-architecture definition happens. The main goal of verification is to ensure functional correctness of the design before the design tape-out. However, with increasing design complexities in the era of advanced node technologies, the scope of verification is also evolving to include much more than functionality, such as verification of performance and power target.

Although simulation of the design RTL (Register Translate Language) remains the primary vehicle for verification, other methodologies, like formal verification (RTL to pre-layout, pre-layout to post-layout, and RTL to post-layout), power-aware simulations (static and dynamic power analysis), emulation/FPGA prototyping, STA, and gate-level simulation (GLS) checks, are used for efficiently verifying all aspects of the design including physical verification before tape-out.

ASIC validation is a process in which the manufactured ASIC design is tested for all functional correctness in a lab setup. This is done using the actual ASIC assembled on a test board, or a reference board, along with all other components of the system for which the chip was designed.

The goal is to validate all use cases of the ASIC that a customer might eventually have in a true deployment and to qualify the design for all these usage models.

© Springer Nature Switzerland AG 2020
K. Golshan, *The Art of Timing Closure*,
https://doi.org/10.1007/978-3-030-49636-4_8

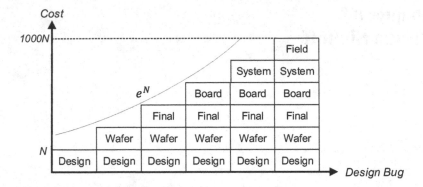

Fig. 8.1 Cost of discovering design bugs

Validation happens initially for individual features and interfaces of the ASIC. It can also be involved in running real software/applications that stress test all the features of the design.

The validation team usually consists of both hardware and software engineers as the overall process involves validating the ASIC in a system-level environment with real software running on the hardware.

Although design validation seems a phase after the silicon process, there are some companies that use the term validation in a broader sense wherein validation is classified as the activities before and after IC availability. Hence, validation is also referred to as pre-silicon validation (indicating activities before the manufactured ASIC is available) and post-silicon validation which is after the ASIC design is manufactured.

It is important to note that the cost to detect and fix an ASIC design bug increases exponentially as the product moves from design to final product delivery. For example, if it costs N units to fix a problem or bug during the design phase, the same problem might cost $1000N$ (this is an arbitrary scale and depends on various organizational business models) to be fixed during field test. Figure 8.1 illustrates the costs at different ASIC design process stages.

8.1 Formal Verification

Formal verification, or Logic Equivalence Check (LEC), refers to a technique that mathematically verifies that two design descriptions being verified are functionally equivalent. LEC provides a formal proof that the output from synthesis and the physical design tools match the original RTL code. All of this is done without having to run a single simulation.

An ASIC design passes through various steps, such as synthesis, physical design, design signoff, ECO, and numerous optimizations, before it reaches production. At every stage, one needs to make sure that the logical functionality is intact and does not break because of any of the automated or manual changes. If the functionality changes at any point during the process, the entire chip becomes faulty. Therefore, LEC is one of the most important checks in the chip process.

In general, LEC is a three-phase process—setup, mapping, and compare.

At the setup phase, the LEC tool reads in two design descriptions—golden and revised. The golden design description could be the design's RTL or the design's synthesized (e.g., pre-layout) netlist. The revised design description is based on the design's routed (e.g., post-layout) netlist.

In addition, the execution of LEC requires a list of libraries (e.g., standard cells and macros) and formal verification constraints. These verification constraints can include constraints such as ignoring specified cells, some scan connections, and input/output pins.

In transition from the setup phase to the mapping phase, both golden and revised design descriptions are flattened and mapped to the key points. The usual key points are:

- Registers (flip-flops and latches)
- Primary inputs and outputs
- Floating signals
- Assignment statements in the revised design description
- Black boxes

During the mapping phase, the LEC tool maps the key points (both golden and revised) and compares them. If there are differences, a report is generated.

In general, there are two types of mapping. One is name-based and the other is non-based mapping. Name-based mapping is used for gate-to-gate mapping such as the pre-layout and post-layout netlists. The non-name-based is useful when golden and revised design descriptions have completely different names (e.g., instance and nets).

The key points that are unmapped during this stage are:

- Extra (the key points that are present in the golden or revised design)
- Unreachable (the key points that are not observable such a primary inputs)
- Non-mapped (the key points that are observable without corresponding instances/ nets)

In the compare phase, the LEC tool compares both golden and revised design descriptions' key points to determine if they are:

- Equivalent
- Nonequivalent
- Inverted equivalent
- Aborted

Figure 8.2 shows the LEC flow.

It is important to note that blocks with critical logical failure found at the time of signoff will cause production delays of the ASIC. At times, the logical connectivity is broken while doing manual fixes or timing ECOs.

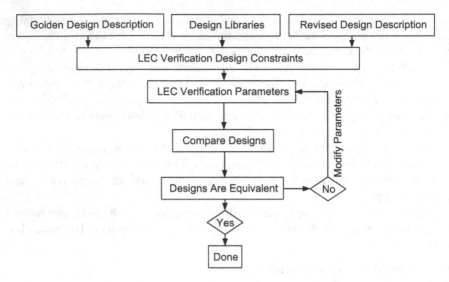

Fig. 8.2 Formal verification (LEC) flow

Since use of MMMC methodology minimizes manual ECOs, one can see that by using MMMC, the formal verification phase is favorably impacted.

8.2   Timing Verification

STA is another important part of ASIC design verification. However, there is an age-old problem that continues to challenge engineers in that STA has some complications with optimization. There is a miscorrelation between implementation timing and signoff timing when the physical design and STA tools use different timing engines.

The only alternative to address this problem is to margin the design during the physical design phase so as not to incur violations during signoff timing which uses golden STA tools. Before performing timing analysis, the extraction files (SPEF) and netlist (Verilog) need to be exported from the physical design tool.

The most common critical paths for timing verifications and optimization are:

- Input to register
- Register to register
- Register to output
- Input to output

Figure 8.3 shows the most common timing paths in a design.

To understand the nature of these timing mismatches between physical design and golden STA tools, one needs to report timing for the same starting and endpoint for a given path that has violations (i.e., setup/hold with negative timing slack) and make sure the violations for both tools are identical.

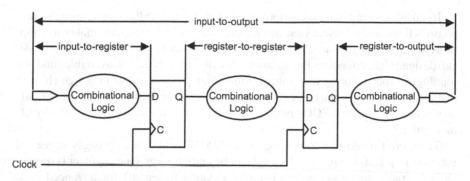

Fig. 8.3 Common timing path illustration

If there are different paths with the same start and endpoint, the conclusion is not valid. If the result is not valid, change uncertainty values (i.e., setup and hold) to compensate for the difference, provided both paths use the same design constraints.

Once physical design and golden STA tools are correlated, if there is any hold and/or setup timing violations, ECOs are applied in order to fix.

With power being one of the most critical design metrics for advanced node technologies, adding excessive gates to ensure minimal violations at signoff is not a proper trade-off. This leads to design power increases and may require the ASIC area be increased. This, in turn, would require the entire physical design flow to be repeated.

The iterations required to close timing differences between physical design and signoff timing are further complicated by the fact that signoff STA timing engines are not physically aware (i.e., typical signoff ECO flow). Placement of inserted buffers, inverters, and upsizing/downsizing cells is left as a post-processing step for ECO generation.

Highly utilized designs inherently have a lack of vacant space. Therefore, the placement of new or ECO cells (e.g., buffers and inverters) can be dramatically different from what is assumed by the optimization algorithms. This results in a significant mismatch between the assumed interconnect parasitic extraction during optimization and the actual placement and routing of the ECO cells in the design.

There are potential uncertainties involved in ECO cell placement which affect assumptions made in a typical signoff ECO flow. For instance, ECO cell placement may impact the timing of paths that have already met timing.

Not only is the impact on timing unknown; it is also impossible to tell which paths will be affected by the movement of ECO cells. In a highly utilized design, the ECO cells may not be placed where they should be placed. This may result in what previously may have been timing clean and could potentially have many violations after routing and placement legalization of the inserted ECO cells.

Therein are just some of the issues in using the typical design flow. In order to address, the MMMC flow is recommended.

It should be noted that timing optimization during the ASIC design timing signoff phase is limited to a logical view of the ASIC design and, therefore, makes assumptions about the placement of inserted cells that vary radically from the actual physical design. Understanding the location of cells, the vacancies available, and the topology of routes during the timing optimization process results in design changes that predictably match the reality of design implementation. This capability not only produces better quality ECOs but also minimizes the impact to paths that already meet timing.

Therefore (in addition to the use of the MMMC flow), it is strongly suggested that both physical design and STA engineer be one engineer. One engineer is responsible for the entire process, from implementation to signoff for advanced node design. Having one physical design engineer for the entire process provides for better results.

Today's design closure requires the physical design tool's timing engine to operate as the same timing engine for the signoff STA tool. In this way, "what-if" analysis can be performed before committing changes to an ECO. Some solutions on the market start with signoff results but lack the integration to perform "what-if" analysis by iterating through the signoff tools. Having all design timing violations fixed during physical design (i.e., after final) routing using MMMC methodology reduces the number of ECO iterations. This allows the last piece of the puzzle to achieve timing closure convergence (i.e., physically aware).

Using physical design tools for timing closures (i.e., physically aware) reduces uncertainty that occurs during legalized placement and routing of an implemented ECO. It's important to note that using golden STA tools (i.e., nonphysically aware) for final design timing signoff should not be used for ECO implementation. It should be used only for checking the quality of results (QoR).

Another consideration is computing power. One of the issues for ASIC design of advanced nodes is the increased physical design and verification run time. For example, the physical and timing analysis engineer must run every mode-corner combination. The quickest way to arrive at a consolidated answer is to distribute the mode-corner combinations across a computer farm and run them in parallel.

There may be some opportunity to bind some operating corners by focusing on extreme corners, but with temperature inversion affecting cell delays and with phase-shift mask uncertainty at 20 nm and below, one can never be sure that the corners have caught all violating paths between the expected best- and worst-case conditions. Today, exhaustive timing analysis and optimization are the recommended approach in order to ensure that a design works under all possible operational conditions.

Once one has distributed all the mode corners to individual servers, the bottleneck becomes the pure processing time required for a single mode corner. Physical design, verification (i.e., timing and physical), and parasitic extraction tools must be architected to leverage multi-core processing.

Today's EDA tool scales reasonably well up to four cores and provide practical scalability up to eight. But while portions of the timing flow today can scale beyond eight, the performance speedup beyond eight has a negligible return when considering the overall timing optimization flow.

Scalability with an increasing number of cores is a must-have for analyzing large advanced ASIC designs, especially with servers containing up to 16 cores. Because scalability is largely dictated by the percentage of processing steps that are multi-threaded, it is not enough to claim that only portions of the timing optimization and extraction flows are threaded.

For example, all steps, from importing the design to reporting or generating output files, must be multi-threaded. This is because the capacity of timing optimization, which is typically in the realm of implementation (e.g., physical design and STA) tools, is generally limited in capacity when optimizing over multiple views.

Most EDA vendors recommend optimizing a subset of mode corners so as not to exceed the capacity limitations of their EDA tools. When optimizing timing across more than 100 mode corners, it is critical to have the capacity to construct a common timing graph that represents a composite picture of all modes and corners within the ASIC design.

Reduction of mode corners must not lead to a loss in accuracy or missed failing endpoints. Otherwise, the yield of timing fixes becomes inaccurate and has a negligible benefit. Miscorrelation between implementation and signoff engines increases the number of ECO iterations needed to close timing of a design. Users have gone through great lengths to build custom solutions that leverage timing attributes and reports from the signoff engine for the purpose of timing optimization.

With today's complex ASIC designs that have many modes and corners, it is imperative to have a common script for generating all STA scripts for all design implementation projects. This ensures the correctness of the STA process.

For example, the *build_sta_tcl* (this *Perl* script is shown in the **Signoff Scripts** section of this chapter) generates the following scripts providing all necessary scripts for STA:

- Project environmental files for each design library (e.g., fast and slow)
- Project constraint files for multiple modes and corners
- *run_sta.tcl* for analyzing design timing for multiple modes and corners
- *sta_setup.env* file for all modes and corners to be sourced by *sta.sh*
- *sta_sh* file to perform STA based on *setup.env* file
- *run_all* STA command file for executing timing analysis

For instance, the following shows an environmental file for project *moonwalk* under /project/moonwalk/implementation/timing/ENV directory. This file contains the path to all specific project-related timing library (i.e., .lib) files for each corner. The following is a sample for best-case corner:

```
### Best-Case Environmental ###
set STDCELL_LIB_FF stdcells_m40c_1p1v_ff
set IOCELL_LIB_FF   io35u_m40c_1p8v_ff
```

```
### Default max cap/trans limits ###
set MAX_CAP_LIMIT          0.350 ;  # 350ff
set MAX_TRANS_LIMIT        0.450  ;  # 450ps

### Setup search path and libraries ###
read_lib [list \
/common/libraries/node20/lib/stdcells/stdcells_m40c_1p1v_ff.lib \
/common/IP/G/node20/pads/lib/io35u_m40c_1p8v_ff.lib \
/common/IP/G/node20/PLL/lib/node20_PLLm40c_1p1v_ff.lib \
/project/moonwalk/implementation/physical/mem/\
RF_52x18/RF_52x18_m40c_1p1v_ff.lib  ]

### Set Delay Calculation and SI Variables ###
set_si_mode -delta_delay_annotation_mode arc \
                    -analysisType aae \
                    -si_reselection delta_delay \
                    -delta_delay_threshold 0.01
  set_delay_cal_mode -engine aae -SIAware true
  set_global  timing_cppr_remove_clock_to_data_crp true
```

Another example is the *run_all* commands that are used to execute STA for all the modes and corners for an ASIC design. This example shows the functional Memory Built-in Self-Test (MBIST), scan capture, and scan shift modes at best-case (MIN) and worst-case (MAX) corners. This can be extended to many modes and corners in the design:

```
STA.sh moonwalk func setup max moonwalk
STA.sh moonwalk func hold min moonwalk
STA.sh moonwalk func hold max moonwalk
STA.sh moonwalk mbist hold min moonwalk
STA.sh moonwalk mbist setup max moonwalk
STA.sh moonwalk func setup max moonwalk
STA.sh moonwalk scanc hold min moonwalk
STA.sh moonwalk scanc hold max moonwalk
STA.sh moonwalk scans hold min moonwalk
```

Standard Delay Format (SDF) for basic gate-level simulation needs to be generated once STA is complete without design timing violations.

If design timing violations occur during basic gate-level simulation, they are usually a result of an incorrect design constraint. In order to close the timing violations, the design constraint needs to be corrected, and STA needs to be redone.

Comprehensive gate-level simulation takes a long time and can continue even after the design has been taped out.

It's a common practice to tape out a design and then start the FEOL (base layers) and stop at *contact* layer. If the comprehensive gate-level simulation shows a bug in the design or more timing violations, one needs to use spare cells from the standard cell and/or metal ECO libraries and perform ECO without breaking the existing design timing.

8.3 Physical Verification

Physical verification is the final phase of any ASIC physical design before submission of the device for fabrication. The main objective of physical verification is to ensure the functionality of ASIC designs and to minimize the risk of creating a design that cannot be manufactured.

Comprehensive physical verification can be an iterative process. It has a time complexity that is linear with respect to the size of the design and sublinear with respect to EDA tools' memory usage. One of today's advanced ASIC design implementation challenges is to reduce the iterative physical design verification process, in terms of the time required, and to be able to improve throughput by means of automating the process of design physical verification.

Advanced nodes have dramatically decreasing device sizes with respect to the number of transistors and routing layers involved in the net result for physical verification of larger layout databases. Because of this, physical verification against the rules is more complex for advanced nodes. Verifying this amount of data puts strain on the software and computer infrastructure.

Therefore, to improve physical verification efficiency and debugging procedures, it is recommended to make sure that the ASIC is physically designed and implemented by employing a correct-by-construction methodology.

In preparation for a smooth physical design process, the following steps need to be performed:

- Export partial layout GDS (cell's abstract and interconnects) from physical design database.
- Import partial physical design GDS in layout database and merge with actual library layout GDS or DFII.
- Export final merged design GDS from layout database.
- Convert post-layout netlist including decap and filler cells to Circuit Design Language (CDL).

One of the most common problems during post-layout to CDL netlist conversion is the mismatch in Spice model names dealing with external CDL netlists such as IP or memories.

Once netlist conversion is done, it is strongly recommended to check the model types in post-layout CDL to make sure they are identical. Here is an inverter CDL example that uses PMOS and NMOS transistor model types. If an external instance, such as memory or IP, uses different model types (e.g., PM and NM transistors) instead of PMOS and NMOS transistors), one needs to edit the final CDL netlist to ensure all model types are identical:

```
$ Model Declaration
.model pmos_name PMOS
.model nmos_name NMOS

$ Inverter Netlist
.subcktinverter in out VDD VSS
mx0 out in VDD VDD  pmos_name w=WIDTH l=Length
mx1 out in VSS  VSS    nmos_name w=WIDTH l=Length
.end
```

Several checks are performed during physical verification:

- Layout Versus Schematic (LVS)
- Design Rule Check (DRC)
- Electrical Rule Check (ERC)

LVS verification examines two electrical circuits (i.e., actual layout and its schematic) to see if they are equivalent with respect to their connectivity and total transistor count. One electrical circuit corresponds to transistor-level schematics or netlist (reference), and another electrical circuit is the result of the extracted layout from the physical database. Through the verification process, if the extracted layout is equivalent to the transistor-level netlist, they should function identically.

One of the primary issues with LVS verification is the repeated iteration of design checking needed to find and remove comparison errors between the layout extracted and the transistor-level netlist (CDL). The cycles involved in the LVS verification consist of GDS data output from the physical database, transistor-level generated netlist, the actual LVS run itself, error diagnosis, and error correction. Thus, one of the objectives during LVS is to keep the length of time needed to complete the verification cycle as minimal as possible.

The most common LVS issues are as follows:

- Shorts: Two or more wires that should not be connected have been and must be separated. The most problematic is power and ground shorts.
- Opens: Wires or components that should be connected are left dangling or only partially connected. These must be connected properly.
- Component Mismatches: Components of an incorrect type have been used (e.g., device-mode transistor type NMOS vs. NM).
- Missing Components: An expected component has been left out of the layout, such as decap cells.
- Parameter Mismatch: Components in the CDL netlist can contain properties. The LVS tool can be configured to compare these properties to a desired tolerance. If the tolerance is not met, then the LVS run will have a property error. In this case, one could set the tolerance value (e.g., 2%).

Currently, there are two ways to improve the LVS verification cycle time. The first consideration must be given to the capacity and performance of the machine infrastructure along with the type of physical verification software. Physical verification software must execute fast and provide accurate results that can be easily traced in case of error. The second consideration is to use hierarchical verification features rather than flat verification.

Hierarchical verification features are used to minimize the amount of data to check and to identify the errors through the usage of hierarchical cells and black box techniques.

It should be pointed out that although the hierarchical verification method is far superior to that of the flat verification, LVS software that incorporates both hierarchical and flat comparisons is far better than verification software that is either strictly hierarchical or flat.

By recognizing the components that simply match between the netlist and layout (such as standard cell libraries), memory block and other intellectual property (IP) elements can be compared in a hierarchical manner while allowing other design elements such as analog blocks and macro cells to maintain a flat representation. In this way, debugging performance is drastically enhanced.

Finally, it is recommended to begin the verification process during an early stage of the design in order to ensure that the physical database is correct. Under the assumption that most of today's physical design tools can produce error-free place and route designs, the most common sources of LVS errors arise during the floorplanning stage and, most often, are related to power and ground connections. Power and ground shorts, and/or opens, impact device (transistor) recognition during LVS verification, which then leads to an extremely long execution time.

DRC is considered a prescription for preparing photomasks that are used in the fabrication of the ASIC design. The major objective of the layout or DRC is to obtain optimal circuit yield without design reliability losses. The more conservative the design rules are, the more likely it leads to correctly manufactured ASIC designs. However, the more aggressive the design rules are, the greater the probability of yield losses.

DRC software uses the so-called DRC decks during the verification process. It is interesting to note that as geometries decrease in size, design rule checking increases. In other words, as semiconductor manufacturing becomes more complex, so do the DRC decks. These complex DRC decks must be composed in an efficient manner. If the DRC rules are not coded in an optimized way in a DRC deck, they tend to require more machine run time and/or machine's memory capacity in order to complete verification.

Another consideration is to use a comprehensive DRC deck during physical verification. If the physical verification is not performed using a comprehensive DRC deck, the result can be low yield or no yield at all.

For a DRC verification to be considered comprehensive, one must make sure the DRC deck checks all the rules correctly and accurately and is able to properly identify and address yield-limiting issues. The most common yield-limiting issues are:

- Charge accumulation due to antenna effects
- Inadequate planarity for multiple layers that are required by CMP
- Metal line mechanical stress
- ESD and latch-up

During the Antenna Rule Check (ARC) verification, ratio calculation in conjunction with wire-charged accumulation is used. In the ratio calculation method, the following ratios can be calculated:

- Wire length to the connecting gate width
- Wire perimeter to the connecting gate area
- Wire area to the connecting gate area

For wire-charged accumulation, assuming N is the metal layer currently to be etched, the following methods can be considered:

- Layer N connecting to the gate
- Layer N plus all layers below forming a path to the gate
- Layer N plus all layers below layer N

Another aspect of DRC allows one to check for DFM (e.g., *contact/via* overlaps and end-of-line enclosures). The DFM rules are considered optional and are provided by the silicon manufacturer. From a yield perspective, it is beneficial to check the ASIC physical design for DFM rule violations and correct the errors as much as possible without influencing the overall die area.

ERC verification is intended to verify an ASIC design electrically. In comparison with LVS that verifies the equivalence between the reference and extracted netlists, ERC checks for electrical errors such as open input pins or conflicting outputs. A design can pass LVS verification but may fail to pass ERC checks.

For example, if there is an unused input in the reference netlist, then the extracted (routed) netlist will also contain the same topology. In this case, the result of the LVS process will be correct by matching both circuits, whereas the same circuits will cause an error during ERC verification (because a floating input gate can lead to excess current leakage).

In the past, ERC verification was used to check the quality of manually captured schematics. Since manual schematic capturing is no longer used for digital designs, ERC verification is mainly performed on the physical data. The ERC verification process is a custom verification rather than a generic verification.

The user can define many electrical rules for the purpose of verification. These rules can be as simple as checking for floating wires or more complex, such as identifying the number of N-well-to-power or P-well-to-substrate contacts, latch-up checks for electrostatic discharge (ESD) in the N-well CMOS process.

In the N-well CMOS process, the bulk of the PMOS is the N-well. It is isolated from the substrate and, thus, can be connected to the source. On the other hand, the bulk of the NMOS is the substrate itself, and, thus, the bulk of the NMOS can't be connected to the source.

8.4 Signoff Scripts

The following *Tcl* script exports Verilog netlists from the physical design tool for timing analysis (without power and ground, gate-level simulation, and physical verification (with power and ground, if the design has different power domains). In addition, it exports DEF file for power and noise analysis.

```
### Setup Environment ###
source /project/implementation/physical/TCL/moonwalk_config.tcl

restoreDesign ../MOONWALK/moonwalk.enc.dat moonwalk
source /project/implementation/physical/TCL/moonwalk_setting.tcl

### Export Netlist Without Power and Ground ###
saveNetlist   /project/implementation/physical/NET/moonwalk_post_
route_npg.v\
             -topCell moonwalk \
             -excludeLeafCell \
             -excludeTopCellPGPort {VDD VSS} \
             -excludeCellInst {artifacts}

### Export Netlist with Power and Ground ###
saveNetlist   /project/implementation/physical/NET/moonwalk_post_
route_pg.v \
             -topCell moonwalk \
             -excludeLeafCell \
             -includePowerGround \
             -includePhysicalCell dioclamp \
             -excludeCellInst {LOGO}

### Export DEF ###
defOut   /project/implementation/physical/DEF/-routing   def/moon-
walk.def

if { [info exists env(FE_EXIT)] && $env(FE_EXIT) == 1 } {exit}
```

The following *Tcl* script is used for netlist-to-netlist and *Tcl*-based manual ECOs:

```
### Setup Environment ###
source /project/implementation/physical/TCL/moonwalk_config.tcl

restoreDesign ../MOONWALK/frt.enc.dat moonwalk
source /project/implementation/physical/TCL/moonwalk_setting.tcl
```

```
generateVias


saveDesign -tcon ../MOONWALK/frt_before_eco.enc -compress


### Netlist-to-Netlist ECO ###
ecoDesign -noEcoPlace -noEcoRoute ../MOONWALK/frt_before_eco.enc.
dat moonwalk \
 ../moonwalk_after_eco.vg
addInst artifacts LOGO
### Place LOGO Instance Back ###
defIn ../../DEF/LOGO.def
ecoPlace

#### TCL Based ECO ###
setEcoMode -honorDontUse false -honorDontTouch false -honorFixed-
Status false
setEcoMode -refinePlace false -updateTiming false -batchMode true

setOptMode -addInstancePrefix ECO_
source ../PNR/eco.tcl
refinePlace -preserveRouting true
checkPlace

### ECO Routing ###
setNanoRouteMode -droutePostRouteLithoRepair false
setNanoRouteMode -routeWithLithoDriven false
generateVias
ecoRoute
routeDesign
saveDesign -tcon ../DESIGNS/frt.enc -compress

setFillerMode -corePrefix FILL -core "List of Filler and de-cap
cells"
addFiller


saveDesign ../MOONWALK/moonwalk.enc -compress


if { [info exists env(FE_EXIT)] && $env(FE_EXIT) == 1 } {exit}
exit
```

The following is a *Perl* script that is used to perform Logic Equivalence Check:

```perl
#!/usr/local/bin/perl
use Time::Local;
$pwd = `pwd`;
chop $pwd;
$mode = 0777;

&parse_command_line;


$Output = $pwd.'/LEC';
unless(-e $Output) { print("    Creating LEC directory...\n");
mkdir($Output,$mode) }

$Target = $pwd.'/LEC/INSTRUCTION';
unless(-e $Target) {
  open(Output,">$Target");
  print(Output "1. Edit env/\$TOP_MODULE_lib.tcl\n");
  print(Output "2. Edit env/\$TOP_MODULE_scan.tcl\n");
  print(Output "3. Edit net/*.f\n");
  print(Output "4. Comment out unnecessary runs in run_all\n");
  print(Output "5. Launch run_all\n");
  print("    Created File: INSTRUCTION\n");
  close (Output);
  chmod 0777,$Target; }

$Output = $pwd.'/LEC/net';
unless(-e $Output) { print("    Creating NET directory...\n");
mkdir($Output,$mode) }

$Target = $pwd.'/LEC/net/'.$TopLevelName.'_rtl_files.f';
open(Output,">$Target");
print(Output "${pwd}/LEC/net/${TopLevelName}_rtl.v");
print("    Created File: ./net/${TopLevelName}_rtl_files.f\n");
close (Output);
chmod 0777,$Target;

$Target = $pwd.'/LEC/net/'.$TopLevelName.'_pre_layout_netlist.f';
open(Output,">$Target");
print(Output "${pwd}/LEC/net/${TopLevelName}_pre.v");
print("    Created File: ./net/${TopLevelName}_pre_layout_
netlist.f\n");
close (Output);
chmod 0777,$Target;

$Target                    =                    $pwd.'/LEC/
```

```
net/'.$TopLevelName.'_post_layout_netlist.f';
open(Output,">$Target");
print(Output "${pwd}/LEC/net/${TopLevelName}_cts.vg");
print("        Created File: ./net/${TopLevelName}_post_layout_
netlist.f\n");
close (Output);
chmod 0777,$Target;

$Output = $pwd.'/LEC/env';
unless(-e $Output) { print("    Creating ENV directory...\n");
mkdir($Output,$mode) }

$Target = $pwd.'/LEC/env/'.$TopLevelName.'_scan.tcl';
open(Output,">$Target");
print(Output "//### Add Scan Conditions to Disable Scan ###\n\n");
print(Output "vpx add pin constraint 0 scan_mode -both\n");
print(Output "vpx add pin constraint 0 scan_enable -both\n");
print("        Created File: ./env/${TopLevelName}_scan.tcl\n");
close (Output);
chmod 0777,$Target;

$Target = $pwd.'/LEC/env/'.$TopLevelName.'_lib.tcl';
open(Output,">$Target");
print(Output  "//### Add ALL Libraries : Standard Cell, IOs,
Memories, Analog macros ###\n\n");
print(Output  "//Include  Actual  Library  Names  with  Full
Paths\n\n");
print(Output "vpx read library -statetable -liberty -both \\\n");
print(Output "    /Pointer to Standard Cell Libraries\\\n");
print(Output "    /Pointer to Input and Output Libraries\\\n");
print(Output "        /Pointer to Memories and Analog Macros
Libraries\n");
print("        Created File: ./env/${TopLevelName}_lib.tcl\n");
close (Output);
chmod 0777,$Target;

 $Target = $pwd.'/LEC/set_up_env.src';
  unless(-e $Target) {
  open(Output,">$Target");
  print(Output "### LEC environments ###\n\n");

  print(Output "setenv PATH ${pwd}\n");
  print(Output "setenv LEC_DIR \$\{PATH\}/LEC\n");
  print(Output "setenv NET_DIR \$\{LEC_DIR\}/net\n");
   print(Output "setenv ENV_TCL_FILE \$\{LEC_DIR\}/env/\$\{TOP_
```

```
MODULE\}_lib.tcl\n");
    print(Output "setenv ENV_SCAN_FILE \$\{LEC_DIR\}/env/\$\{TOP_
MODULE\}_scan.tcl\n");
   print(Output "setenv Pointer to LEC Tool\n");
   print("   Created File: ./set_up_env.src\n");
   close (Output);
   chmod 0777,$Target; }

$Target = $pwd.'/LEC/LEC.sh';
unless(-e $Target) {
   open(Output,">$Target");
   print(Output "#!/bin/csh  -f\n");
   print(Output "# \$Id: \$\n\n");
   print(Output "set thisscript = \$\{0\}\n");
   print(Output "echo \"@@ Running \$\{thisscript\}...\"\n\n");
   print(Output "## Display Usage\n");
   print(Output "if ( (\$#argv < 3) || (\$#argv > 3) ) then\n");
   print(Output " echo \"\"\n");
   print(Output " echo \"Err\"\"or: Missing Argument(s)\"\n");
   print(Output " echo \"Usage:   > \$thisscript:t <module_name>
<release_tag>\"\n");
    print(Output "  echo \"Example: > \$thisscript:t lhr_digital
r1234\"\n");
   print(Output " echo \"\"\n");
   print(Output " exit\n");
   print(Output "endif\n\n");
   print(Output "setenv  TOP_MODULE    \$1 \;\n");
   print(Output "shift\n");
   print(Output "setenv  MODE          \$1 \;\n");
   print(Output "shift\n");
   print(Output "setenv  LABEL         \$1 \;\n");
   print(Output "shift\n\n");
   print(Output "source set_up_env.src\n\n");
   print(Output "setenv IMP \$LEC_DIR\n\n");
   print(Output "setenv RTL_FILES \$\{TOP_MODULE\}_rtl_files\n");

print(Output  "setenv  PRE_NET_FILE  \$\{TOP_MODULE\}_pre_layout_
netlist\n");

print(Output  "setenv  PST_NET_FILE  \$\{TOP_MODULE\}_post_layout_
netlist\n\n");
 print(Output "setenv RTL_FILE \$\{NET_DIR\}/\$\{RTL_FILES\}.f\n");
                         print(Output            "setenv
PRE_NET \$\{NET_DIR\}/\$\{PRE_NET_FILE\}.f\n");
```

```
print(Output          "setenv          PST_NET           \$\{NET_DIR\}/\$\
{PST_NET_FILE\}.f\n\n");
 print(Output "setenv TOP_DIR      \$\{IMP\}/\$\{TOP_MODULE\}\n");
 print(Output "setenv CURRENT_REV \$\{TOP_DIR\}/\$\{LABEL\}\n\n");
 print(Output "if (! -d \$\{TOP_MODULE\}) then\n");
 print(Output "       echo \"@@ Creating \$\{TOP_MODULE\}\"\n");
 print(Output "         mkdir -p \$\{IMP\}/ \$\{TOP_MODULE\}\n");
 print(Output "endif\n\n");
 print(Output "if (! -d \$\{CURRENT_REV\}) then\n");
    print(Output   "                     echo  \"@@  Creating   \$\
{CURRENT_REV\}\"\n");
 print(Output "       mkdir -p \$\{CURRENT_REV\}\n");
 print(Output "endif\n\n");
 print(Output "if (! -d \$\{CURRENT_REV\}/rpt) then\n");
 print(Output "  echo \"@@ Creating \$\{CURRENT_REV\}/rpt\"\n");
 print(Output "  mkdir -p \$\{CURRENT_REV\}/rpt\n");
 print(Output "endif\n\n");
 print(Output "if (! -d \$\{CURRENT_REV\}/log) then\n");
 print(Output "  echo \"@@ Creating \$\{CURRENT_REV\}/log\"\n");
 print(Output "  mkdir -p \$\{CURRENT_REV\}/log\n");
 print(Output "endif\n\n");
 print(Output " lec -nogui -xl -64 \$\{IMP\}/run_LEC.tcl |& tee \
 \$\{CURRENT_REV\}/log/\$\{TOP_MODULE\}_run.log\n");
 print("   Created File: ./LEC.sh\n");
 close (Output);
 chmod 0777,$Target; }

$Target = $pwd.'/lec/run_LEC.tcl';
unless(-e $Target) {
 open(Output,">$Target");
 print(Output "tclmode\n\n");
 print(Output "vpx set undefined cell black_box\n\n");
 print(Output "vpx set undriven signal 0 -golden\n\n");
    print(Output  "//vpx  set  analyze  option  -auto  -ANALYZE_
ABORT -effort high\n\n");
 print(Output "vpx dofile \$env(ENV_DO_FILE)\n\n");

 print(Output "// Read Pre-layout and Post-layout Netlist\n\n");
 print(Output "if \{\$env(MODE) == \"pre2pst\"\} \{\n");
 print(Output " vpx read design -verilog2k -golden -sensitive -file
\$env(PRE_NET)\n");
 print(Output " vpx read design -verilog -revised -sensitive -file
\$env(PST_NET)\}\n\n");

 print(Output "// Read RTL file & Pre-layout\n\n");
```

```
   print(Output "if \{\$env(MODE) == \"rtl2pst\"\} \{\n");
  print(Output " vpx read design -verilog2k -golden -sensitive -file
\$env(RTL_FILE)\n");
  print(Output " vpx read design -verilog -revised -sensitive -file
\$env(PRE_NET)\}\n\n");

  print(Output "// Read RTL file and Post-layout Netlist\n\n");
  print(Output "if \{\$env(MODE) == \"rtl2pst\"\} \{\n");
  print(Output " vpx read design -verilog2k -golden -sensitive -file
\$env(RTL_FILE)\n");
  print(Output " vpx read design -verilog -revised -sensitive -file
\$env(PST_NET)\}\n\n");

  print(Output "vpx set root module \$env(TOP_MODULE) -both\n\n");
  print(Output "vpx uniq -all -nolibrary -summary -gol\n\n");
  print(Output "//vpx set mapping method -name first -unreach\n");
  print(Output "//vpx set flatten model -gated_clock -hrc_verbose\
      -NODFF_TO_DLAT_FEEDBACK      -nodff_to_dlat_zero   -noout_to_
inout -noin_to_inout\n");
  print(Output "//vpx set datapath option -auto -merge\n\n");
    print(Output "vpx  set  flatten  model  -gated_clock  -seq_
constant -seq_constant_feedback\n\n");

  print(Output "vpx set compare option -allgenlatch\n");
  print(Output "vpx set compare option -noallgenlatch\n");
  print(Output "vpx set cpu limit 36 -hours -walltime -nokill\n");
  print(Output "vpx set compare options -threads 2\n\n");
  print(Output "vpx report black box -nohidden\n\n");

  print(Output "//Disable Scan Mode \n\n");
  print(Output "vpx source \$env(ENV_SCAN_FILE)\n\n");
  print(Output "vpx set system mode lec\n\n");

    print(Output   "vpx   remodel   -seq_constant   -seq_constant_
feedback -verbose\n\n");
  print(Output "vpx map key points\n\n");
  print(Output "vpx add compare points -all\n\n");
  print(Output "vpx analyze datapath -verbose -merge\n\n");

  print(Output "// Effort Options: low, super, ultra, complete\n");
  print(Output "vpx set compare effort super\n\n");
  print(Output "vpx compare\n\n");

  print(Output "vpx report compare data -noneq \
  > \$env(IMP)/\$env(TOP_MODULE)/\$env(LABEL)/rpt/noneq_pre2pst.
```

```
rpt\n");
  print(Output "vpx report unmapped points \
      >    \$env(IMP)/\$env(TOP_MODULE)/\$env(LABEL)/rpt/unmapped_
pre2pst.rpt\n");
  print(Output "\n");
  print(Output "vpx save session -replace \
  \$env(IMP)/\$env(TOP_MODULE)/\$env(LABEL)/session\n\n");

  print(Output "vpx exit -f\n");
  print("   Created File: ./run_LEC.tcl\n");
  close(Output);
  chmod 0777,$Target; }

$Target = $pwd.'/LEC/run_all';
if(-e $Target) {
  open(Output,">>$Target");
  print("   Appended to File: ./run_all\n"); }
else {
  open(Output,">$Target");
  print("   Created File: ./run_all\n"); }

print(Output "LEC.sh $TopLevelName pre2pst PRE2PST\n");
print(Output "LEC.sh $TopLevelName rtl2pre RTL2PRE\n");
print(Output "LEC.sh $TopLevelName rtl2pst RTL2PST\n");
close(Output);
chmod 0777,$Target;
print("   Done\n");

sub parse_command_line {
    for($i=0; $i<=$#ARGV; $i++){
      $_ = $ARGV[$i];
      if(/^-t$/){$TopLevelName=$ARGV[++$i]}
      if (/^-h\b/) { &print_usage }
      }

    unless($TopLevelName){print "ERROR: incorrectly specified com-
mand line. \
    Use -h for more information.\n";
    exit(0);}}

sub print_usage {
  print"\nusage: build_lec_tcl -t TopLevelName\n";
  print"\n";
  print"   -t   # Where is TopLevelName of the chip or block\n";
  print"\n";
```

```
        print" This    command    creates    LEC   directory    under
project/implementation/timing/VER \
  for setup files  to run LEC.\n"}
```

The following is *Perl* script that generates the following files for performing STA. These files are:

- Constraints.tcl
- Setup_env.src (source file)
- Run_STA.tcl
- STA.sh (Static Timing Analysis shell file)

```perl
#!/usr/local/bin/perl
use Time::Local;
$pwd = `pwd`;
chop $pwd;
$mode = 0777;
&parse_command_line;
@pwd = split(/\//,$pwd);
$implementaion_dir= '';
$physical_dir=';

for($i=0; $i<=3; $i++){
  $implementaion_dir= $implementaion_dir.$pwd[$i].'/'; }
  $timing_dir = $ implementaion_dir.'timing';
  $physical_dir = $implementaion_dir.'physical';
  $Output = $timing_dir.'/STA';
unless(-e  $Output)  {  print("Creating  STA  directory...\n");
mkdir($Output,$mode) }

$Target = $timing_dir.'/STA/INSTRUCTION';
unless(-e $Target) {
  open(Output,">$Target");
  print(Output "1. Edit ENV/*.env and add files for additional cor-
ners if necessary\n");
  print(Output "2. Edit STA/run_STA.tcl to add or change analysis
corners as needed\n");
  print(Output "3. Edit TCL/*_constraints.tcl\n");
  print(Output "4. Edit TCL/read_*_netlist_spef.tcl for process
and hierarchy if needed\n");
  print(Output "5. Comment out unnecessary runs in run_all and add
runs as needed\n");
  print(Output "6. Launch STA/run_all\n");
  print("  Created File: INSTRUCTION\n");
  close (Output);
```

```
   chmod 0777,$Target; }

$Output = $timing_dir.'/NET';
unless(-e $Output) { print("   Creating NET directory...\n");
mkdir($Output,$mode) }

$Target = $timing_dir.'/NET/'.$TopLevelName.'_frt.vg';
$Source = $physical_dir.'/NET/'.$TopLevelName.'_frt.vg';
symlink($Source,$Target);
print(" Created Link: ./NET/${TopLevelName}_frt.vg\n");
chmod 0777,$Target;

$Output = $timing_dir.'/SPEF';
unless(-e $Output) { print("   Creating SPEF directory...\n");
mkdir($Output,$mode) }

$Target = $timing_dir.'/SPEF/'.$TopLevelName.'_max.spef.gz';
$Source = $physical_dir.'/SPEF/'.$TopLevelName.'_max.spef.gz';
symlink($Source,$Target);
print(" Created Link: ./SPEF/${TopLevelName}_max.spef.gz\n");
chmod 0777,$Target;

$Target = $timing_dir.'/SPEF/'.$TopLevelName.'_min.spef.gz';
$Source = $physical_dir.'/SPEF/'.$TopLevelName.'_min.spef.gz';
symlink($Source,$Target);
print(" Created Link: ./SPEF/${TopLevelName}_min.spef.gz\n");
chmod 0777,$Target;
$Output = $timing_dir .'/ENV';
unless(-e  $Output)  {  print(" Creating  ENV  directory...\n");
mkdir($Output,$mode) }

$Target = $timing_dir.'/ENV/node20_sta_ff.env';
unless(-e $Target) { open(Output,">$Target");

 print(Output "### Display Script Entry Message ###\n");
  print(Output "set thisscr \"node20_sta_ff\"\n");
  print(Output "puts \"@@ Entering \$\{thisscr\}...\"\n\n");
  print(Output "set STDCELL_LIB_FF  stdcells_m40c_1p1v_ff\n");
  print(Output "set IOCELL_LIB_FF    io35u_m40c_1p1v_ff\n\n");

  print(Output "### Default max cap/trans limits ###\n\n");
  print(Output "set MAX_CAP_LIMIT         0.350 \; # 350ff\n");
  print(Output "set MAX_TRANS_LIMIT    0.450 \;  # 450ps\n\n");

  print(Output "### Setup search path and libraries ###\n");
```

```
  print(Output "read_lib [list \\\n");
   print(Output "           /common/libraries/node20/lib/stdcells/std-
cells_m40c_1p1v_ff.lib \\\n");
   print(Output "      /common/IP/G/node20/pads/lib/io35u_m40c_1p1v_
ff.lib \\\n");
    print(Output "           /common/IP/G/node20/PLL/node20_m40c_1p1v_
PLL_ff.lib/ \\\n");
   print(Output "      $physical_dir/MEM/RF_52x18/RF_52x18_m40c_1p1v_
ff.lib \\\n"); ]\n\n");

  print(Output "### Set Delay Calculation and SI Variables ###\n");
    print(Output "set_si_mode  -delta_delay_annotation_mode  arc
\\\n");
   print(Output "                              -analysisType aae \\\n");
   print(Output "                              -si_reselection delta_delay
\\\n");
   print(Output "                              -delta_delay_threshold 0.01
\n\n");
   print(Output "set_delay_cal_mode -engine aae -SIAware true\n");
    print(Output "set_global  timing_cppr_remove_clock_to_data_crp
true\n");
   print("     Created File: ./ENV/node20_sta_ff.env\n");
   close (Output);
   chmod 0777,$Target;
}
$Target = $timing_dir.'/ENV/node20_sta_ss.env';
unless(-e $Target) {open(Output,">$Target");
$Target = $timing_dir.'/ENV/node20_sta_ff.env';
unless(-e $Target) { open(Output,">$Target");

print(Output "### Display Script Entry Message ###\n");
print(Output "set thisscr \"node20_sta_ss\"\n");
print(Output "puts \"@@ Entering \$\{thisscr\}...\"\n\n");
print(Output "set STDCELL_LIB_SS  stdcells_125c_1p05v_ss\n");
print(Output "set IOCELL_LIB_SS    io35u_125c_1p5v_ss\n\n");

print(Output "### Default max cap/trans limits ###\n\n");
print(Output "set MAX_CAP_LIMIT        0.350 \; # 350ff\n");
print(Output "set MAX_TRANS_LIMIT  0.450 \; # 450ps\n\n");

  print(Output "### Setup search path and libraries ###\n");
  print(Output "read_lib [list \\\n");
   print(Output "           /common/libraries/node20/lib/stdcells/std-
cells_ff_125c_1p05v.lib \\\n");
  print(Output "      /common/IP/G/node20/pads/lib/io35u_125c_1p05v_
```

```
ss.lib \\\n");
 print(Output "      /common/IP/G/node20/PLL/node20_PLL_125c_1p05v_
ss.lib/ \\\n");
 print(Output "    $physical_dir/MEM/RF_52x18/RF_52x18_125c_1p05v_
ss.lib \\\n"); ]\n\n");

  print(Output "### Set Delay Calculation and SI Variables ###\n");
    print(Output  "set_si_mode  -delta_delay_annotation_mode   arc
\\\n");
  print(Output "                         -analysisType aae \\\n");
  print(Output "                         -si_reselection delta_delay
\\\n");
   print(Output "                          -delta_delay_threshold
0.01\n\n");
  print(Output "set_delay_cal_mode -engine aae -SIAware true\n");
  print(Output "set_global   timing_cppr_remove_clock_to_data_crp
true\n");
  print(" Created File: ./ENV/node20_sta_ss.env\n");
  close (Output);
  chmod 0777,$Target;
}

$Output = $timing_dir.'/TCL';
unless(-e $Output)  {  print(" Creating  TCL  directory...\n");
mkdir($Output,$mode) }
$Target  =  $timing_dir.'/STA/TCL/'.$TopLevelName.'_constraints.
tcl';
open(Output,">$Target");
print(Output "if \{\$FUNC_TYPE == \"func\"\} \{\n");
print(Output " echo \"### Source Functional Constraints ###\"\n");
print(Output " set_case_analysis 0 [get_pins top_level/scan_mux/
SCAN]\n");
print(Output "    source  \$STA/TCL/\$\{TOP_MODULE\}_func_cons.
tcl\n");
print(Output "\}\n\n");

print(Output "if \{\$FUNC_TYPE == \"scans\"\} \{\n");
print(Output " echo \"### Source Scan Shift Constraints ###\"\n");
print(Output " set_case_analysis 1 [get_pins top_level/scan_mux/
SCAN]\n");
print(Output " set_case_analysis 1 [get_pins top_level/scan_mode_
buff/Y\n");
print(Output "    source  \$STA/TCL/\$\{TOP_MODULE\}_scans_cons.
tcl\n");
print(Output "\}\n\n");
```

```
print(Output "if \{\$FUNC_TYPE == \"scanc\"\} \{\n")
print(Output "   echo \"### Source Scan Capture Constraints
*****\"\n");
print(Output " set_case_analysis 1 [get_pins top_level/scan_mux/
SCAN]\n");
print(Output " set_case_analysis 0 [get_pins top_level/scan_mode_
buff/Y\n");
print(Output "   source \$STA/TCL/\$\{TOP_MODULE\}_scanc_cons.
tcl\n");
print(Output "\}\n\n");
print(" Created File: ./TCL/${TopLevelName}_constraints.tcl\n");
close (Output);
chmod 0777,$Target;


$Target = $timing_dir.'/TCL/read_'.$TopLevelName.'_netlist_spef.
tcl';
open(Output,">$Target");
print(Output "setDesignMode -process 20nm \n\n");
print(Output "read_verilog   \"\$WORKDIR/NET/${TopLevelName}_frt.
vg\"\n\n");
print(Output   "set_top_module  $TopLevelName  -ignore_undefined_
cell\n\n");
print(Output "read_spef   \"\$WORKDIR/SPEF/${TopLevelName}_\
                 \$\{MAXMIN\}.spef.gz \"\n\n");
print(Output "report_annotated_parasitics -list_not_annotated > \
\$\{OUT_DIR\}/log/parasitics_command_\$\{MAXMIN\}.log\n");
print(" Created File:  ./TCL/read_${TopLevelName}_netlist_spef.
tcl\n");
close (Output);
chmod 0777,$Target;


$Target = $timing_dir.'/STA/set_up_env.src';
unless(-e $Target) {
  open(Output,">$Target");
  print(Output "##  STA environments ###\n\n");
  print(Output "setenv WORKDIRROOT       $implementaion_dir \n");
  print(Output "setenv WORKDIR              \$\{WORKDIRROOT\}/\n");
  print(Output "setenv STA                      \$\WORKDIR\n");
                print(Output         "setenv        PHYSICAL
$physical_dir\n");
  print("   Created File: ./set_up_env.src\n");
  close (Output);
  chmod 0777,$Target;
}
```

```
$Target = $timing_dir.'/STA/STA.sh';
unless(-e $Target) {
  open(Output,">$Target");
  print(Output "source set_up_env.src\n\n");

print(Output "\\rm -rf .clock_group_default* .STA_emulate_view_
de*\n\n");
  print(Output "set thisscript = \$\{0\}\n");
  print(Output "echo \"@@ Running \$\{thisscript\}...\"\n\n");
  print(Output "setenv USER STA\n");
  print(Output "echo    \$USER\n\n");

  print(Output "###  Display Usage ###\n");
  print(Output "if ( (\$#argv < 5) || (\$#argv > 10) ) then\n");
  print(Output "  echo \"\"\n");
  print(Output "  echo \"Err\"\"or: Missing Argument(s)\"\n");
  print(Output "  echo \"Usage:   > \$thisscript:t <module_name>
<func_type>\
                      <analysis_type> <lib_opcond> <label> [local]
[power] [outdir <out_dir>]\"\n");
  print(Output "  echo \"Example: > \$thisscript:t coral scanc
setup max Freeze1\"\n");
  print(Output "  echo \"\"\n");
  print(Output "  exit\n");
  print(Output "endif\n\n");
  print(Output "setenv  TOP_MODULE    \$1 \;\n");
  print(Output "shift\n");
  print(Output "setenv  FUNC_TYPE     \$1 \;\n");
  print(Output "shift\n");
  print(Output "setenv  ANALYSIS_TYPE \$1 \;\n");
  print(Output "shift\n");
  print(Output "setenv  LIB_OPCOND    \$1 \;\n");
  print(Output "shift\n");
  print(Output "setenv  LABEL  \$1 \;   #Label used to save sta
data\n");
  print(Output "shift\n\n");
  print(Output "setenv SCR_ODIR \
    "\$WORKDIRROOT/\$USER/\$\{TOP_MODULE\}/\$\{LABEL\}/\
    \$\{FUNC_TYPE\}_\$\{ANALYSIS_TYPE\}_\$\{LIB_OPCOND\}\"\n");
  print(Output "setenv LOCAL_ODIR \
    "\$STA/\$\{TOP_MODULE\}/\$\{FUNC_TYPE\}_\$\{ANALYSIS_TYPE\}_\
    \$\{LIB_OPCOND\}_\$\{LABEL\}\"\n\n");
  print(Output "setenv OUT_DIR \$\{SCR_ODIR\} \; # Default =
Scratch Drive\n\n");
  print(Output "while (\$#argv)\n");
```

```
print(Output "switch ( \$1 )\n");
print(Output " case \"local\"\n");
print(Output " setenv OUT_DIR \"\$\{LOCAL_ODIR\}\"\n");
print(Output "    breaksw\n");
print(Output "    case \"outdir\"\n");
print(Output "    setenv OUT_DIR \"\$\{2\}\"\n");
print(Output "    shift\n");
print(Output "    breaksw\n");
print(Output "    default:\n");
print(Output "    echo \"@@ Sorry you \
  entered an unknown switch or missing space : \$1\"\n");
print(Output "       exit\n");
print(Output "        breaksw\n");
print(Output "  endsw\n");
print(Output "  shift\n");
print(Output "end \; # Command Line option processing\n\n");
print(Output "if (! -d \$\{OUT_DIR\}/rpt) then\n");
 print(Output "  echo \"@@ Creating ./directory/sub-directory:
\$\{OUT_DIR\}/rpt\"\n");
 print(Output "  mkdir -p \$\{OUT_DIR\}/rpt\n");
 print(Output "endif\n");
 print(Output "\n");
 print(Output "if (! -d \$\{OUT_DIR\}/) then\n");
 print(Output "  echo \"@@ Creating ./directory/sub-directory:
\$\{OUT_DIR\}/sdc\"\n");
 print(Output "  mkdir -p \$\{OUT_DIR\}/SDC\n");
 print(Output "endif\n");
 print(Output "\n");
 print(Output "if (! -d \$\{OUT_DIR\}/SDF) then\n");
 print(Output "  echo \"@@ Creating ./directory/sub-directory:
\$\{OUT_DIR\}/sdf\"\n");
 print(Output "  mkdir -p \$\{OUT_DIR\}/SDF\n");
 print(Output "endif\n");
 print(Output "\n");
 print(Output "if (! -d \$\{OUT_DIR\}/session) then\n");
 print(Output "  echo \"@@ Creating ./directory/sub-directory:
\$\{OUT_DIR\}/session\"\n");
 print(Output "  mkdir -p \$\{OUT_DIR\}/session\n");
 print(Output "endif\n");
 print(Output "\n");
 print(Output "if (! -d \$\{OUT_DIR\}/log) then\n");
 print(Output "  echo \"@@ Creating ./directory/sub-directory:
\$\{OUT_DIR\}/log\"\n");
 print(Output "  mkdir -p \$\{OUT_DIR\}/log\n");
 print(Output "endif\n");
```

```
  print(Output "\n");
  print(Output "if (! -d \$\{OUT_DIR\}/lib) then\n");
   print(Output "   echo \"@@ Creating ./directory/sub-directory:
\$\{OUT_DIR\}/lib\"\n");
  print(Output "  mkdir -p \$\{OUT_DIR\}/lib\n");
  print(Output "endif\n");
  print(Output "\n");
  print(Output "STA -init \$\{STA\}/run_STA.tcl \\\n");
  print(Output "     -log \
            $\{OUT_DIR\}/log/\$\{TOP_MODULE\}_\$\{FUNC_TYPE\}_\$\
{ANALYSIS_TYPE\}_\
  $\{LIB_OPCOND\}_STA.log -nologv \\\n");
  print(Output "     -cmd \
            $\{OUT_DIR\}/log/\$\{TOP_MODULE\}_\$\{FUNC_TYPE\}_\$\
{ANALYSIS_TYPE\}_\
  $\{LIB_OPCOND\}.cmd |& tee\n");
  print(Output "\n");
  print(Output "chmod -R 777 \$\{OUT_DIR\}/*\n");
  print("   Created File: ./STA.sh\n");
  close (Output);
  chmod 0777,$Target;
}

$Target = $pwd.'/STA/run_STA.tcl';
unless(-e $Target) {
  open(Output,">$Target");
  print(Output "### Setup STA ###\n\n");
  print(Output "date\n");
  print(Output "\n");
  print(Output "set WORKDIRROOT    [getenv WORKDIRROOT]\n");
  print(Output "set WORKDIR              [getenv WORKDIR]\n");
  print(Output "set STA                     [getenv STA]\n");
  print(Output "set LIB_OPCOND       [getenv LIB_OPCOND]\n");
  print(Output "set TOP_MODULE       [getenv TOP_MODULE]\n");
  print(Output "set FUNC_TYPE         [getenv FUNC_TYPE]\n");
  print(Output "set ANALYSIS_TYPE   [getenv ANALYSIS_TYPE]\n");
  print(Output "set LABEL                  [getenv LABEL]\n");
  print(Output "set OUT_DIR              [getenv OUT_DIR]\n");
  print(Output "\n");

  print(Output "### STA Common Settings ###\n\n");
  print(Output "\n");
  print(Output "set timing_report_group_based_mode true\n");
  print(Output "set report_timing_format\
                      {instance arc cell delay incr_delay load
```

```
arrival required}\n");
  print(Output "set_analysis_mode -analysisType onChipVariation -cppr
both\n");
  print(Output "\n");

  print(Output "### Analysis Corner Settings ###\n\n");
  print(Output "\n");
  print(Output "if \{\$LIB_OPCOND == \"min\"\} \{\n");
  print(Output "  source \$STA/ENV/node20_sta_ff.env\n");
  print(Output "  set MAXMIN  min\n");
  print(Output "\}\n");
  print(Output "\n");
  print(Output "if \{\$LIB_OPCOND == \"max\"\} \{\n");
  print(Output "  source \$STA/ENV/node20_sta_ss.env\n");
  print(Output "  set MAXMIN  max\n");
  print(Output "\}\n");
  print(Output "\n");

  print(Output "### Read Design and Parasitic and Netlist ###\n\n");
  print(Output "\n");
    print(Output   "source    \$WORKDIR/TCL/read_\$\{TOP_MODULE\}_
netlist_spef.tcl -verbose\n");
  print(Output "\n");

  print(Output "if \{\$LIB_OPCOND == \"min\"\} \{\n");
print(Output " set_analysis_mode -analysisType onChipVariation\n");
  print(Output "  setOpCond BEST -library node20_sta_ff\n");
  print(Output "  set CORNER \"ff\"\n");
  print(Output "  if \{\$ANALYSIS_TYPE == \"hold\"\} \{\n");
  print(Output "    set_timing_derate -early 1.0 -clock\n");
  print(Output "    set_timing_derate -late  1.1 -clock\n");
  print(Output "  \}\n");
  print(Output "\}\n");
  print(Output "\n");

  print(Output "if \{\$LIB_OPCOND == \"max\"\} \{\n");
print(Output " set_analysis_mode -analysisType onChipVariation\n");
  print(Output "  setOpCond WORST -library node20_sta_ss\n");
  print(Output "  set CORNER \"ss\"\n");
  print(Output "  if \{\$ANALYSIS_TYPE == \"hold\"\} \{\n");
  print(Output "    set_timing_derate -early 0.95 -clock\n");
  print(Output "    set_timing_derate -late  1.0  -clock\n");
  print(Output "  \}\n");
  print(Output "\}\n");
  print(Output "\n");
```

```
print(Output "if \{\$ANALYSIS_TYPE == \"hold\"\} \{\n");
print(Output "   if \{\$LIB_OPCOND == \"max\"\} \{\n");
print(Output "        set DELAYTYPE min\n");
print(Output "   \} else \{\n");
print(Output "        set DELAYTYPE \$LIB_OPCOND\n");
print(Output "   \}\n");
print(Output "\} else \{\n");
print(Output "   set DELAYTYPE \$LIB_OPCOND\n");
print(Output "\}\n");
print(Output "\n");

print(Output "### Apply Constraints ###\n\n");
print(Output "\n");
   print(Output "source \$STA/TCL/\$\{TOP_MODULE\}_constraints.
tcl -verbose\n");
print(Output "\n");
print(Output "if \{\$MAXMIN == \"max\"\} \{\n");

print(Output    "         set_clock_uncertainty   -setup   0.300
[all_clocks]\n");

print(Output    "         set_clock_uncertainty   -hold    0.300
[all_clocks]\n");
   print(Output "\}\n");
   print(Output "\n");
   print(Output "if \{\$MAXMIN == \"min\"\} \{\n");

print(Output    "         set_clock_uncertainty   -setup   0.300
[all_clocks]\n");

print(Output    "         set_clock_uncertainty   -hold    0.150
[all_clocks]\n");
   print(Output "\}\n");
   print(Output "\n");
   print(Output "set_multi_cpu_usage -localCpu 4\n");
   print(Output "\n");
   print(Output "set_propagated_clock [all_clocks ]\n");
   print(Output "\n");
   print(Output "if \{\$\{FUNC_TYPE\} == \"mbist\"\} \{\n");
   print(Output " set_false_path -from [all_inputs]\n");
   print(Output " set_false_path -to [all_outputs]\n");
   print(Output "\}\n");
   print(Output "date\n");
   print(Output "\n");
   print(Output "update_timing\n");
```

```
print(Output "date\n\n");
print(Output "save_design   \
  \$\{OUT_DIR\}/session/\$\{TOP_MODULE\}_\$\{FUNC_TYPE\}_\
  \$\{ANALYSIS_TYPE\}_\$\{LIB_OPCOND\}_session -overwrite\n");
print(Output "date\n");

print(Output "### Report Timing ###\n\n")
print(Output "#set report_default_significant_digits 3\n");
print(Output "report_constraint -all_violators   > \
  \$\{OUT_DIR\}/RPT/\$\{TOP_MODULE\}_\$\{FUNC_TYPE\}_\
  \$\{ANALYSIS_TYPE\}_\$\{LIB_OPCOND\}.allvio.rpt\n");
print(Output "\n");
print(Output "report_constraint -all_violators  -verbose > \
  \$\{OUT_DIR\}/RPT/\$\{TOP_MODULE\}_\$\{FUNC_TYPE\}_\
  \$\{ANALYSIS_TYPE\}_\$\{LIB_OPCOND\}.allvio_verbose.rpt\n");
print(Output "\n");
print(Output "if \{[string equal \$ANALYSIS_TYPE hold]\} \{\n");
print(Output "    # 10 path report full_clock\n");
print(Output "    report_timing -path_type full_clock  -early\
                          -from [all_registers -clock_pins] -to
[all_registers -data_pins] \
                          -net -max_paths 10 -max_slack 0 \\\n");

print(Output "      > \
  \$\{OUT_DIR\}/RPT/\$\{TOP_MODULE\}_\$\{FUNC_TYPE\}_\
  \$\{ANALYSIS_TYPE\}_\$\{LIB_OPCOND\}.tim.reg_reg.rpt\n");
  print(Output "    # 3 path report no clock path\n");
  print(Output "    report_timing -early -from [all_registers -clock_
pins] \
                          -to [all_registers -data_pins]  -net -max_
paths 3 -max_slack 0 \\\n");
  print(Output "      > \
  \$\{OUT_DIR\}/RPT/\$\{TOP_MODULE\}_\$\{FUNC_TYPE\}_\
  \$\{ANALYSIS_TYPE\}_\$\{LIB_OPCOND\}.tim.reg_reg_short.rpt\n");
  print(Output "\n");
  print(Output "\} else \{\n");
  print(Output "    # 10 path report full_clock\n");
  print(Output "    report_timing -path_type full_clock \
                          -late -from [all_registers -clock_pins]
-to [all_registers -data_pins]\
                          -net -max_paths 10 -max_slack 0 \\\n");
  print(Output "      > \
  \$\{OUT_DIR\}/RPT/\$\{TOP_MODULE\}_\$\{FUNC_TYPE\}_\
  \$\{ANALYSIS_TYPE\}_\$\{LIB_OPCOND\}.tim.reg_reg.rpt\n");
  print(Output "    # 3 path report no clock path\n");
```

```
  print(Output "    report_timing -early -from [all_registers -clock_
pins] \
                          -to [all_registers -data_pins]  -net -max_
paths 3 -max_slack 0  \\\n");
  print(Output "      > \
  \$\{OUT_DIR\}/RPT/\$\{TOP_MODULE\}_\$\{FUNC_TYPE\}_\
  \$\{ANALYSIS_TYPE\}_\$\{LIB_OPCOND\}.tim.reg_reg_short.rpt\n");
  print(Output "\}\n\n");
  print(Output "#report_timing \
                          -path full_clock_expanded  -delay min_
max -cap -tran -crosstalk_delta \\\n");
  print(Output "# -net -max_paths 30 -slack_lesser_than 0 \\\n");
  print(Output "# > \
  \$\{OUT_DIR\}/RPT/\$\{TOP_MODULE\}_\$\{FUNC_TYPE\}_\
  \$\{ANALYSIS_TYPE\}_\$\{LIB_OPCOND\}.tim.allvio.rpt\n");
  print(Output "\n");
  print(Output "#report_timing -path full_clock_expanded \
                          -delay min_max -cap -tran -cross-
talk_delta \\\n");
  print(Output "# -net -max_paths 3  \\\n");
  print(Output "# > \
  \$\{OUT_DIR\}/RPT/\$\{TOP_MODULE\}_\$\{FUNC_TYPE\}_\
  \$\{ANALYSIS_TYPE\}_\$\{LIB_OPCOND\}.tim.all.rpt\n");
  print(Output "\n");

 print(Output "check_timing -verbose    > \
 \$\{OUT_DIR\}/RPT/\$\{TOP_MODULE\}_\$\{FUNC_TYPE\}_\
 \$\{ANALYSIS_TYPE\}_\$\{LIB_OPCOND\}.check_timing.rpt\n");
 print(Output "report_analysis_coverage > \
 \$\{OUT_DIR\}/RPT/\$\{TOP_MODULE\}_\$\{FUNC_TYPE\}_\
 \$\{ANALYSIS_TYPE\}_\$\{LIB_OPCOND\}.analysis_coverage.rpt\n");
  print(Output "report_min_pulse_width   > \
 \$\{OUT_DIR\}/RPT/\$\{TOP_MODULE\}_\$\{FUNC_TYPE\}_\
 \$\{ANALYSIS_TYPE\}_\$\{LIB_OPCOND\}.min_pulse_width.rpt\n");
 print(Output "report_clocks     > \
 \$\{OUT_DIR\}/RPT/\$\{TOP_MODULE\}_\
     \$\{FUNC_TYPE\}_\$\{ANALYSIS_TYPE\}_\$\{LIB_OPCOND\}.clocks.
rpt\n");
 print(Output "\n");

 print(Output "report_case_analysis > \
 \$\{OUT_DIR\}/RPT/\$\{TOP_MODULE\}_\$\{FUNC_TYPE\}_\
 \$\{ANALYSIS_TYPE\}_\$\{LIB_OPCOND\}.case_analysis.rpt\n");
  print(Output "report_clock_timing -type skew  >> \
 \$\{OUT_DIR\}/RPT/\$\{TOP_MODULE\}_\$\{FUNC_TYPE\}_\
```

```
 \$\{ANALYSIS_TYPE\}_\$\{LIB_OPCOND\}.clocks.rpt\n");
  print(Output "report_clock_timing -type summary  >> \
   \$\{OUT_DIR\}/RPT/\$\{TOP_MODULE\}_\$\{FUNC_TYPE\}_\
  \$\{ANALYSIS_TYPE\}_\$\{LIB_OPCOND\}.clocks.rpt\n");
  print(Output "report_clock_timing -type latency -verbose >> \
   \$\{OUT_DIR\}/RPT/\$\{TOP_MODULE\}_\$\{FUNC_TYPE\}_\
  \$\{ANALYSIS_TYPE\}_\$\{LIB_OPCOND\}.clocks.rpt\n");
  print(Output "\n");
   print(Output "if \{\$\{FUNC_TYPE\} == \"func\" || \$\{FUNC_
TYPE\} == \"func1\" || \
   \$\{FUNC_TYPE\} == \"scans\" || \$\{FUNC_TYPE\} == \"scanc\"\} \
{\n");
  print(Output "#write_sdf \$\{OUT_DIR\}/SDF/\$\{TOP_MODULE\}_\$\
{FUNC_TYPE\}_\
   \$\{ANALYSIS_TYPE\}_\$\{LIB_OPCOND\}.sdf \\\n");
  print(Output "# -precision 4 -version 3.0 -setuphold split -rec-
rem split \
   -edges noedge -condelse \n");
  print(Output "\n");
  print(Output "# write_sdf \$\{OUT_DIR\}/SDF/\$\{TOP_MODULE\}_\$\
{FUNC_TYPE\}_\
   \$\{ANALYSIS_TYPE\}_\$\{LIB_OPCOND\}.sdf \\\n");
   print(Output "#  -precision 4 -version 3.0 -edges noedge -con-
delse \n");
  print(Output "\n");
   print(Output "write_sdf \$\{OUT_DIR\}/SDF/\$\{TOP_MODULE\}_\$\
{FUNC_TYPE\}_\
    \$\{ANALYSIS_TYPE\}_\$\{LIB_OPCOND\}_nocondelse.sdf \\\n");
   print(Output " -precision 4 -version 3.0 -edges noedge -min_
period_edges none\n");
  print(Output "\n");
  print(Output "write_sdc \$\{OUT_DIR\}/SDC/\$\{TOP_MODULE\}_sta.
sdc\}\n");
  print(Output "\n");
  print(Output "if \{\$\{FUNC_TYPE\} == \"func\"\} \{\n");
  print(Output "  set_false_path -from [all_inputs]\n");
  print(Output "  set_false_path -to [all_outputs]\n");
  print(Output "  date\n");
  print(Output "\n");
  print(Output "  update_timing\n");
  print(Output "  date\n");
  print(Output "\n");
  print(Output "  report_constraint -all_violators   \\\n");
  print(Output "    > \
   \$\{OUT_DIR\}/RPT/\$\{TOP_MODULE\}_\$\{FUNC_TYPE\}_\
```

```
  \$\{ANALYSIS_TYPE\}_\$\{LIB_OPCOND\}_IO_Falsed.allvio.rpt\n");
  print(Output "\n");
   print(Output "    report_constraint -all_violators    -verbose
\\\n");
  print(Output "    > \
  \$\{OUT_DIR\}/RPT/\$\{TOP_MODULE\}_\$\{FUNC_TYPE\}_\
   \$\{ANALYSIS_TYPE\}_\$\{LIB_OPCOND\}_IO_Falsed.allvio_verbose.
rpt\n");
  print(Output "\n");

print(Output "   save_design      \$\{OUT_DIR\}/session/\$\{TOP_
MODULE\}_\
  \$\{FUNC_TYPE\}_\
                \$\{ANALYSIS_TYPE\}_\$\{LIB_OPCOND\}_IO_Falsed_
session -rc -overwrite\n");
  print(Output " date \}\n");
  print(Output "\n");
  print(Output "#report_constraint -all_violators    \\\n");
    print(Output "#  >  \$\{OUT_DIR\}/RPT/\$\{TOP_MODULE\}_\$\
{FUNC_TYPE\}_\
  \$\{ANALYSIS_TYPE\}_\$\{LIB_OPCOND\}.allvio_inout.rpt\n");
  print(Output "\n");
  print(Output "#report_timing -path full_clock_expanded \
  -delay \$DELAYTYPE -from [all_inputs] -to [all_registers -data_
pins]  \\\n");
   print(Output "# -net -cap -tran -crosstalk_delta -max_paths
10 -slack_lesser_than 0 \\\n");
  print(Output "# > \
  \$\{OUT_DIR\}/RPT/\$\{TOP_MODULE\}_\$\{FUNC_TYPE\}_\
  \$\{ANALYSIS_TYPE\}_\$\{LIB_OPCOND\}.tim.ip_reg.rpt\n");
  print(Output "\n");
  print(Output "#report_timing -path full_clock_expanded -delay\
  \$DELAYTYPE -from [all_registers -clock_pins] -to [all_outputs]
\\\n");
   print(Output "# -net -cap -tran -crosstalk_delta -max_paths
10 -slack_lesser_than 0 \\\n");
    print(Output "#  >  \$\{OUT_DIR\}/RPT/\$\{TOP_MODULE\}_\$\
{FUNC_TYPE\}_\
  \$\{ANALYSIS_TYPE\}_\$\{LIB_OPCOND\}.tim.reg_op.rpt\n");
  print(Output "\n");
  print(Output "#report_timing -path full_clock_expanded -delay\
  \$DELAYTYPE -from [all_inputs] -to [all_outputs] \\\n");
   print(Output "# -net -cap -tran -crosstalk_delta -max_paths
10 -slack_lesser_than 0 \\\n");
```

```
    print(Output "# > \
      \$\{OUT_DIR\}/RPT/\$\{TOP_MODULE\}_\$\{FUNC_TYPE\}_\
      \$\{ANALYSIS_TYPE\}_\$\{LIB_OPCOND\}.tim.ip_op.rpt\n");
    print(Output "\n");

    print(Output "exit\n");
    print("   Created File: ./run_STA.tcl\n");
    close(Output);
    chmod 0777,$Target; }
$Target = $pwd.'/STA/run_all';
if(-e $Target) {
  open(Output,">>$Target");
  print("   Appended to File: ./run_all\n"); }
else {
  open(Output,">$Target");
  print("   Created File: ./run_all\n");
}
print(Output "STA.sh $TopLevelName func setup max $TopLevelName\n");
print(Output "STA.sh $TopLevelName func hold min $TopLevelName\n");
print(Output "STA.sh $TopLevelName func hold max $TopLevelName\n");
print(Output "STA.sh $TopLevelName mbist hold min $TopLevelName\n");
print(Output "STA.sh $TopLevelName mbist setup max $TopLevelName\n");
print(Output "STA.sh $TopLevelName scanc hold min $TopLevelName\n");
print(Output "STA.sh $TopLevelName scanc hold max $TopLevelName\n");
print(Output "STA.sh $TopLevelName scans hold min $TopLevelName\n");
close(Output);
chmod 0777,$Target;
print("   Done\n");

sub parse_command_line {
    for($i=0; $i<=$#ARGV; $i++){
      $_ = $ARGV[$i];
      if(/^-t$/){$TopLevelName=$ARGV[++$i]}
      if (/^-h\b/) { &print_usage } }

    unless($TopLevelName){print "ERROR: incorrectly specified com-
mand line. \
              Use -h for more information.\n";
    exit(0);}}

sub print_usage {
    print"\nusage: build_sta_tcl -t TopLevelName\n";
    print"\n";
    print"  -t   # Where TopLevelName is the top-level name of the
```

```
chip or block\n";
    print"\n";
    print"  This command creates a STA directory and setup files
under TIMING directory.\n"}
```

8.5 Summary

Chapter 8 discusses the ASIC signoff process from timing, logical, and physical verification point of views. Integrating signoff and implementation tools has been discussed for many years, and there have been attempts to make it happen through today's EDA technology. One of the key aspects for making the signoff process, especially in the area of timing closures and extraction works, as it should be is to have an EDA system that is built with integration in mind.

This includes common timing engines and unified databases so that the exchange of data from implementation to extraction to timing is a seamless process and results are consistent. Patching together long-standing point tools is generally a suboptimal solution that neither delivers on efficiency or run time performance. Fortunately for the user, there are timing closure solutions on the market that address the items shown above.

The design (physical and timing) engineers should expect their EDA vendors to provide solutions that address today's signoff timing closure deficiencies in the overall design flow process. Otherwise, with the movement to advanced nodes, this phase in the design flow could become the longest step in taping out a design by having to apply many unnecessary design ECOs.

Although EDA vendors are trying to provide solutions that address today's signoff timing closure deficiencies in the overall design flow process, there may be some miscorrelation between physical and static timing engines (e.g., one may be more optimistic or pessimistic with respect to setup and hold time calculation). One could adjust the design margin for either setup or hold, especially on the optimistic timing engine, in order to correlate both timing engines as close as possible.

In this chapter, physical verifications such as Design Rule Checking (DRC), Layout Versus Schematics (LVS), Antenna Design Checking (ARC), and Electrical Rule Checking (ERC) are discussed. Common practices providing an efficient physical design process are also included.

Bibliography

1. J. Lienig, M. *Thiele, Introduction. Fundamentals of Electromigration Aware Integrated Circuit Design* (Springer international, 2018) AG
2. J. Lienig, M. Thiele, *The pressing need for electromigration-aware physical design, in Proceedings of the International Symposium on Physical Design (ISPD)* (March 2018) Monterey, California, USA

3. N. Dershowitz, *Verification: Theory and Practice, Lecture Notes In Computer Science Series* (Springer, Berlin, 2004.

4. Hardware Verification Group, *Digital Logic Synthesis and Equivalence Checking Tools* (Department of Electrical and Computer Engineering, Concordia University, Montreal, Canada, May 2010)

Index

© Springer Nature Switzerland AG 2020
K. Golshan, *The Art of Timing Closure*,
https://doi.org/10.1007/978-3-030-49636-4

Printed in the United States
by Baker & Taylor Publisher Services

Printed in the United States
by Baker & Taylor Publisher Services